People, Forests and Natural Resources
No. 11

General Editors

Klaus Seeland
Franz Schmithüsen

Chair of Forest Policy and Forest Economics
Institute for Human-Environment-Systems
Swiss Federal Institute of Technology (ETH), Zurich

People, Forests and Natural Resources Series

A number of volumes of scientific account of the forest life of tribal groups in South and South-East Asia have so far been published in the series *Man and Forest*. As forests are a relevant but complemented by other renewable natural resources, the scope of the series has been extended to *natural resources*. The anthropologic term "Man" was exchanged against "People" as worldwide trends in citizen's participation and democratic mobilisation mark a change towards political and resource management activism in which the empowerment of people is in focus. Thus the series "People, Forests and Natural Resources" will continue to emphasise research on forests, soil, water, air and other renewable natural resources and publish accounts of how they matter for human society and culture.

This series welcomes contributions covering relevant issues in the people, forest and natural resources interface.

Local Initiatives
Collective Water Management in Rural Bangladesh

To my daughter

Maya

Local Initiatives

Collective Water Management in Rural Bangladesh

by

Jennifer E. Duyne

D.K. Printworld (P) Ltd.
New Delhi

Cataloging in Publication Data — DK
Courtesy: D.K. Agencies (P) Ltd. <docinfo@dkagencies.com>

Duyne, Jennifer E., 1961-

Local initiatives: collective water management in rural Bangladesh / by Jennifer E. Duyne.

p. 23 cm.

(People, forests and natural resources ; no. 11).

Includes bibliographical references (p.)

Includes index.

ISBN 8124602662

1. Water — supply — Bangladesh — Management. 2. Water resources development — Bangladesh. 3. Water resources development — Bangladesh — Citizen participation. I. Title. II. Series: People, forests and natural resources ; no. 11.

DDC 363.610 954 92 21

ISBN 81-246-0266-2
First published in India in 2004

Photographs by : Jennifer E. Duyne and Dieuke Joosten
Drawings by : Dieuke Joosten
Cover Design by : Lukas M. Pleij

Published and printed by:
D.K. Printworld (P) Ltd.
Regd. office : '*Sri Kunj*', F-52, Bali Nagar
New Delhi - 110 015
Phones : (011) 2545-3975; 2546-6019
Fax : (011) 2546-5926
E-mail: dkprintworld@vsnl.net
Web: www.dkprintworld.com

Foreword

DR. Duyne's investigation of local water management practices in rural Bangladesh widens the perspective of relevant indigenous knowledge and endogenous lifestyles that have ever since been a major concern of this series. Water management through local self-help activities facilitated by traditional social structures and solidarity among village communities and local institutions is a significant domain of villagers who are initiating environmental, social and economic sustainability by caring for the basic prerequisites of their livelihood.

Bangladesh is one of the least developed countries in the world and the knowledge of water management as well as the struggle against floods is a legend since centuries. Local knowledge is an invaluable resource in itself and local initiatives to manage this vital resource have saved many village communities from becoming clients of the state. However, detailed accounts of this knowledge, the techniques and skills that have been developed over the ages have rarely attracted the attention of Western scientists as far as Bangladesh is concerned. The author's intensive studies have now contributed to fill this gap in a substantial way. They introduce the reader to a little known world of collective action in rural communities that in most societies has become a task of state administration. As long as these communities organise themselves as local institutions and take their vital interests into their own hands they will manage to overcome many obstacles to their economic development.

This book is essential reading for the concerned general reader and provides professional experts with inspiring accounts of a fascinating world of traditional technologies and organisational practices.

Klaus Seeland
Franz Schmithüsen

Preface

THIS book comes following a deep study I conducted while I was working as a social scientist for the Systems Rehabilitation Project (SRP), a Dutch-funded technical assistance programme with the Bangladesh Water Development Board (BWDB), the leading government agency in charge of surface water management. My main task was to set-up a system to backstop, monitor and evaluate newly established farmers' organisations, which were expected to assume a major role in the operation and maintenance of state provided hydraulic infrastructure.

However, I soon learned that international approaches towards farmers' participation in irrigation management were not directly applicable to the socio-technical context of Bangladesh. In order to come up with some viable policy recommendations, it was necessary to know more about how people in rural Bangladesh themselves manage their water resources. In 1997, during my last year with SRP, I was offered the opportunity to contribute the revision of the Bangladesh Ministry of Water Resources' *Guidelines for People's Participation in Water Development Projects* with a one-year research project on endogenous water management practices in rural Bangladesh. The study constituted an independent component of a broader research project on sustainable water resource management funded by the Government of the Netherlands that was carried out under the auspices of the Centre for Environment, University of Leeds. Most data presented in this study were collected within the framework of that project.

There are many people and institutions I would like to thank for their support to this endeavour. First of all, I would like to thank SRP's socio-economic team, who assisted in fieldwork and to whom I owe much of the insights on local water management practices in rural Bangladesh: Eakub Ali Khan, Moinuddin Tazim, Zainal Abedin, Harun-ur-Rashid Patwari, Saiful Alam, Shakawat Hossain, Mohsiur Rahman and Anjit Kumar Chakrabarti. Without their profound knowledge of people's water management practices gained over many years of fieldwork in rural Bangladesh, and their endurance and enthusiasm in working with me on this research project this book would never have come to existence.

I would like to thank the BWDB and the various international development agencies that supported SRP, for being persuaded about the importance of such a study. I am particularly grateful to the Netherlands Directorate General for Development Cooperation (DGIS) for having funded this research project. I am also grateful to Dr. Shamsul Huda, the then Secretary of the Ministry of Water Resources, for having appreciated the main findings of this study and for expressing his commitment that they would indeed influence the still ongoing water sector reform programme.

Among my friends and colleagues at SRP, I would like to thank in particular Dieuke Josten, associate expert in water resource management, for the many photographs and the drawings documenting people's initiatives in water management. I owe to my colleague Jan Bron, team leader of SRP's Netherlands Technical Assistance Programme, many interesting professional discussions and much of the technical insight I gained over the time on water management.

Many friends and colleagues have supported me in various ways to undertake the endeavour of writing up this study in its present form. Dr. Kamal Siddique, Senior Government Secretary and respected social scientist, encouraged me to dedicate more time to the analysis of my data; Euroconsult, my former employer, allowed me to take leave and ensured me that I would be welcomed back, whenever I would be ready for it. I am grateful to Prof. H.-P. Müller for having welcomed me back at the Department of Social Anthropology of the University of Zurich, thus offering me a stimulating academic environment and much encouragement during the tedious and solitary period of writing-up this book.

I would like to give my warmest thanks to my friend Dr. Suzanne Hanchett, social anthropologist, for the critical comments and for proof reading the first draft of this book.

I am greatful to Prof. Dr. Franz Schmithüsen and Prof. Dr. Klaus Seeland for their interest in indigenous water management practices in Bangladesh and for offering me the opportunity to publish this book within the series on "People, Forests and Natural Resources."

I am grateful to my husband Dr. Jorge Barenstein for having given me his day-to-day intellectual and emotional support. I thank my beloved daughter Maya for accepting with mature endurance my limited availability and for decorating my office with flowers and colourful drawings.

Finally, with immense gratitude, I would like to dedicate this study to my real teachers in water management, the many women and men I met in rural Bangladesh, who gave me much of their time, who trusted me with their information, and who gave me with their extraordinary courage and resilience a valuable lesson of life. I do hope that somehow, some day, my modest role as a mediator between them and the macro-level policy makers may contribute to close the gap between them.

Jennifer E. Duyne

Contents

Abbreviations, Unit & Measures and Currency

BADC	Bangladesh Agricultural Development Corporation
BBS	Bangladesh Bureau of Statistics
BMDA	Barind Multipurpose Development Authority
BWDB	Bangladesh Water Development Board
CIDA	Canadian International Development Agency
DPHE	Department of Public Health and Engineering
DTW	Deep Tubewell
EMG	Embankment Maintenance Group
FAP	Flood Action Plan
FCD	Flood Control and Drainage
FCDI	Flood Control, Drainage and Irrigation
FFW	Food for Work
GOB	Government of Bangladesh
GOEP	Government of East Pakistan
HYV	High Yielding Variety
IFAD	International Fund for Agricultural Development
LGED	Local Government Engineering Department
LLP	Low Lift Pump
MOWR	Ministry of Water Resources
NGO	Non-Government Organisation

O&M	Operation and Maintenance
SDE	Sub-divisional Engineer
SRP	Systems Rehabilitation Project
SSFCDIP	Small Scale Flood Control, Drainage and Irrigation Project
STW	Shallow Tubewell
TA	Technical Assistance
Tk	Taka (Bangladesh currency)
TNO	Thana Nirbahi Officer
UP	Upazila Parishad
WUO	Water Users Organisation

Units & Measures

1 *crore*	10 million
1 *lakh*	100,000
1 *maund*	around 37 kg
1 *seer*	0.9 kg
1 *acre*	100 decimals
1 *bigha*	0.33 acres
1 *kani*	40 decimals
1 *keyar*	0.33 acres

Currency

1 Taka 100 Paisha

1 US Dollar around 45 Tk (1998)

List of Figures, Maps and Tables

Figures and Maps

Tables

List of Visuals

1

Context, Objectives and Approach

Objectives

THIS book is about how people in rural Bangladesh, without any institutional or material support, manage their water resources to make their environment safer or more productive. It shows that Bangladesh's people have an underestimated capacity to organise and develop their own strategies to solve the problems they face; they are neither passive victims of their environment nor merely responding or reacting to programmes or services provided by 'outside' agencies.

The data in this book are based on extensive studies conducted in four regions of Bangladesh. This approach made it possible to uncover significant regional differences in terms of type, frequency and complexity of local water management practices. This book aims at explaining these differences, with the objective to draw some light on the conditions under which people organise to pursue a common goal.

This study was carried out within the framework of a project funded by the Government of the Netherlands between 1990 and 1998. The project aimed at supporting the Water Development Board (BWDB), Bangladesh's leading government agency in charge of surface water resource management to adopt a participatory water management approach by organising water users groups. These groups were expected to assume a major role in the operation and maintenance of state provided hydraulic infrastructure.

The approach to be followed was clearly inspired by international strategies to enhance farmers' participation in irrigation management. Whether this approach would be viable in the specific context of Bangladesh is a question that was not addressed by the project concept, though according to much social research in Bangladesh there are reasons to question people's incentives to participate to the provision of collective goods. In fact, much research on rural Bangladesh reaches the conclusion that the local power structure is an impediment for rural development in general and for participatory collective action in particular. Further, according to several progressive NGOs, social scientists and environmental activists, people in rural Bangladesh do not benefit from BWDB's hydraulic infrastructure, which would further contribute to their lack of motivation to participate to its operation and maintenance.

This book shows, however, that even though formal initiatives to organise water users organisations to enhance people's participation in water management projects were thus far not very successful, people in rural Bangladesh — regardless of their socio-economic status — generally appreciate state provided hydraulic infrastructure and are willing to contribute to its operation and maintenance. Many cases indicate that people regularly take upon themselves the burden to improve hydraulic systems (or to reduce their negative impact) that have to be fine-tuned to local requirements and conditions. People appear to have the material and organisational capacity to repair defective structures, such as breached embankments or leaking regulators, though not necessarily following engineering standards. The rural power structure inevitably defines how public hydraulic infrastructure is managed but this does not necessarily lead to inequitable outcomes, for there is no clear correlation between a stakeholder's socio-economic status and his water management interests. There are cases of conflicts but also of consensus among vast and apparently heterogeneous groups of people.

The book further discusses the important role of 'informal' hydraulic infrastructure, that is, water management systems constructed by the people themselves, which do not figure in any maps and which in spite of their vital importance have so far simply been overlooked. Case studies of over seven hundred local initiatives indicate that people have detailed knowledge and the capacity to design, construct, maintain and operate relatively complex water management systems. They show that people are able to mobilise on their own and make an optimal use of large amounts of human, material and financial resources. However, as will be shown, the capacity to organise is much higher in some regions than in others. This book aims at explaining under what conditions people in rural Bangladesh get together to pursue common goals by analysing the research findings against the backdrop of academic debates on participation, collective action and common property resource management.

Local Initiatives in Water Resource Management

Local initiatives are defined as individual or collective actions, taken without any support or encouragement from external agencies, to control common surface water resources. These endogenous initiatives may relate to traditional indigenous water management practices or they may be triggered by external events, such as new opportunities or constraints associated with the construction of state-provided hydraulic infrastructure. Local initiatives may be taken by informal or semi-formal organisations existing and functioning beyond the observed action or by individuals or groups of people who organise exclusively for that specific purpose. They may benefit the community as a whole or one particular individual or category of stakeholders. They may have a positive, a negative or no impact on other individuals or groups of people. Typical examples of local initiatives relate to the construction of crossdams and drainage outlets or to embankment cuts and also to people's endeavours to take over the management of hydraulic infrastructure constructed by external agencies, infrastructure which is not being operated and maintained according to specific local requirements.

Not all water management initiatives described in this book are undertaken by groups of people. In fact, what is considered relevant is not so much whether a water management system is provided, operated or maintained individually or collectively, but whether the initiative has a broader public impact. Indeed, some of the most costly and sophisticated locally constructed water management infrastructures were provided by private individuals with the means and the power to encroach on and capture public resources. This refers particularly to the infrastructure constructed by salt and shrimp producers in the coastal area, infrastructure having an impact on large numbers of people. The research does not cover water management initiatives by private individuals if these have no immediate impact on other people.

Research Issues and Methodology

Bangladesh's socio-economic conditions have been researched by many national and international social scientists. The research agenda has been strongly influenced by the massive international aid directed at the country since its independence. Accordingly, much research focuses on 'impacts of' or 'responses to' external interventions, leaving little room to find out how people act on their own. A basic distinction may be made between qualitative studies focusing on the village level and macro-level studies working with aggregate quantitative data. Whereas micro-level qualitative studies for their lack of representativeness are often considered of limited relevance for policy-making purposes, macro-level studies tend to be excessively deductive. As they rely exclusively on data gathered through surveys by means of structured individual interviews, they generally fail to uncover issues beyond those which are already, at least intuitively, known or predictable.

This book is primarily based on empirical investigations by a team of six social scientists and two agricultural scientists. The study combined qualitative and quantitative research methods and covered areas sufficiently large and diverse to claim representativeness. During the first research phase (June-August 1997) the objectives were: (a) to identify and collect some basic data about all water management infrastructure constructed, operated or maintained — apparently without any external assistance — by local people, and (b) to find out how and by whom in the absence of any effective formal authority in charge of their O&M — state provided hydraulic systems were being managed. By means of individual and group interviews data were collected about the nature and function of the infrastructure, its main beneficiaries, key actors involved in its construction, the number of voluntary or paid labour-days involved, the type and value of materials used, the financial resources involved, the strategies pursued to mobilise the required resources and about any eventually negative impacts or related conflicts.

The objective of the second research phase (September-December 1997) was to understand the social organisation of local water management initiatives, patterns of resource mobilisation and their appreciation and legitimacy from the perspective of men and women of different socio-economic statuses and interests. To this aim detailed case studies were carried out for each type of initiative. Depending on the number of people involved or affected by an initiative, in-depth interviews were held either with a stratified sample or with the whole group of people involved. Semi-structured interviews were also held with women and men not directly involved in the initiatives in order to find out their perceptions and attitudes towards locally constructed water management infrastructure.

The third and last research phase (January-April 1998) focused on contextual analysis, which was considered instrumental to understand regional differences in number, size and complexity of local water management practices. With the aim to explain the factors that contribute or inhibit the emergence of local initiatives, we selected five representative villages in each research area in which we gathered a variety of data related to local socio-economic and agro-ecological conditions. In order to assess the awareness and response of officials, we also carried out interviews with Union Parishad chairmen and BWDB engineers. Further, we carried out detailed studies on specific issues related to local water management practices. These included, for example, a study of Chaptir Haor's village funds and a study of salt and shrimp production in Banskhali. Over six hundred open-ended interviews were carried out in each research site during a period of about ten months.

Many local water management practices, though logical and legitimate from a local perspective, are formally illegal. To protect our informants, the names of the people and in some cases, of the villages involved in a particular initiative are either fictitious or not mentioned.

Geographical Setting

This book is based on fieldwork in the areas affected by four flood control and drainage schemes that were constructed by the BWDB (or its predecessor WAPDA) between 1966 and 1987. An effort was made to find four representative FCD schemes from distinct regions of Bangladesh. We selected four areas that we knew well. In these areas we had played a key role in ensuring that the rehabilitation and improvement works of BWDB's infrastructure would be based on people's needs and recommendations, an activity that had brought us in close and prolonged contact with the local population. We thus were known and trusted by many people and knew that we could count on the active participation of the local population.

Beel Singri in the north-western district of Chapai-Nawabganj was selected to represent an FCD scheme in a *beel* environment. It refers to an area of land of around 135 km^2 which constitutes the main pillar of the livelihood of over 45,000 people. BWDB's infrastructure consists of four regulators and a 2.7 km embankment that was constructed in 1987. Prior to the construction of the highway-*cum*-embankment along the Mahananda River and the FCD scheme the area was vulnerable to river floods.

Table 1.1: Research areas at a glance

Site	*District*	*Thana*	*No. of Unions*	*No. of Villages*	*Population*	*Area (ha)*
Beel Singri	Chapai-Nawabganj	Nawabganj Sadar and Nachol	3	28	45,153	13,543
Chaptir Haor	Sunamganj	Derai	3	35	30,952	5,061
Polder 64/1b	Chittagong	Banskhali	12	36	254,844	7,740
Polder 55/1	Patuakhali	Galachipa	4	63	77,531	10,600
TOTAL	**4**	**5**	**22**	**162**	**408,480**	**36,944**

Source: GOB/BBS, 1991: Bangladesh Population Census and EGIS (1998).

Chaptir Haor in the north-eastern district of Sunamganj, represents a typical FCD scheme in the *haor* basin, Bangladesh's largest inland depression. Chaptir Haor consists of a saucer-shaped interfluvial area of about 50 km^2 vulnerable to pre-monsoon flash floods. BWDB provided Chaptir Haor with a submersible embankment of 42.5 km in 1977. This type of embankment is only intended to delay the inundation, allowing farmers to harvest their crops before the whole *haor* is submersed in deep water for a period of about six months.

Polder 55/1 is the designation given by the BWDB to an area of land of about 100 km^2 located in the south-western coastal belt, which prior to polderization used to be a *char*. This research area, referred to as Galachipa (the name of the *upazila* covered by the FCD scheme), includes 63 villages belonging to Patuakhali District. BWDB's coastal embankment that surrounds the area was completed in 1977, in order to protect the area against saline water intrusion, tidal flooding and cyclone surges. The embankment is equipped with 35 flushing sluices to facilitate drainage.

Finally, ***Polder 64/1b*** refers to a narrow strip of land, 53.5 km by 2-5 km, located between the Bay of Bengal and the Chittagong Hill Tracts, about 35 km south of Chittagong, the country's second most important urban centre. It covers most of Banskhali Upazila, and accordingly will be referred to as Banskhali. BWDB's embankment, running parallel to the coast, aims to protect the area from saline water intrusion at high tides and from storm surge floods. To facilitate drainage, the embankment is provided with 36 sluices and six culverts. More details about the socio-economic and agro-ecological conditions of these regions will be given in Chapters 6 and 7.

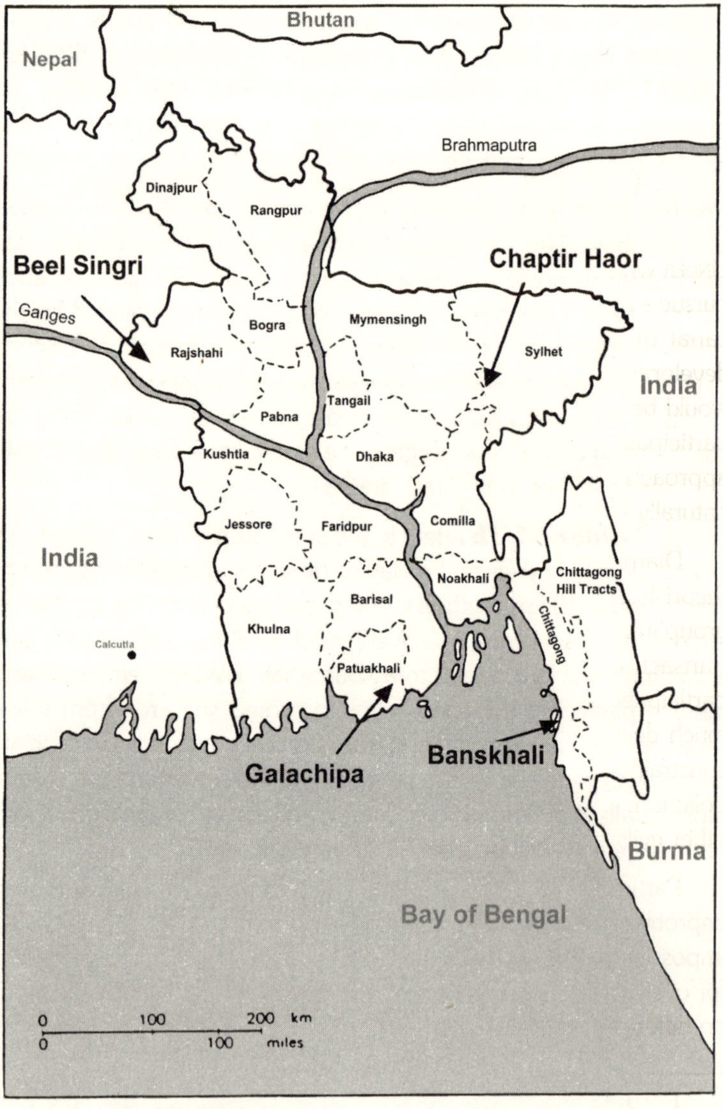

***Fig.* 1.1: Map of Bangladesh with research areas location**

2

Conditions for Collective Action

UNDER what conditions will a group of people voluntarily organise to pursue a common goal, such as building an embankment, an irrigation canal or a drainage system? Many advocates of participatory development and international development agencies assume that it would be sufficient to persuade policy-makers about the benefits of participation, introduce related guidelines, incorporate a participatory approach into a project design and stakeholders' participation will naturally occur.

Diametrically opposed views prevail among collective action theorists, who consider the conflict between individual interests and group interests, people's inclination towards free riding and the high transaction costs[1] of collective action to be serious obstacles to participation. These factors would explain why many public goods much desired by groups of people often are not provided. Similar constraints also would determine the depletion of natural resources, pollution, low citizen participation in democratic elections and many other evils of industrial and developing societies.

Participation optimists, who consider the formation of groups as unproblematic and collective action pessimists, who consider it virtually impossible, have co-existed and systematically ignored one another for over three decades. Both continue to exercise a strong influence on different circles of policy-makers and development practitioners.

---

1. Transaction costs refer to the costs related to the organisation of collective action. They include search costs (identifying possibilities of cooperation), bargaining costs (agreeing on one scheme of cooperation), and monitoring and enforcement costs (Taylor and Singleton 1993: 196).

On the one hand, a high degree of optimism prevails among governmental and non-governmental development agencies, which consider 'participation' through the formation of community organisations a key ingredient for the success of any development project. On the other hand, in particular in the domain of natural resource management, collective action pessimism still legitimises policy pressures either to privatise common property resources or to keep them under central government control.

Increasingly influential, at least in the domain of common property resource management, are the more differentiated perspectives of a number of scholars who recognise the existence of comparatively successful cases of collective action and emphasise the importance of analysing the natural, socio-economic, institutional and political contexts in which they occur.

This chapter first provides a summary of the concepts of participatory development prevailing among collective action optimism. This will be followed by a brief discussion of the most popular paradigms of collective action pessimism.

Collective Action Optimism

> *Participation of users in managing and maintaining water facilities brings many benefits. Participation in planning, operating and maintaining irrigation. . . increases the likelihood that these will be maintained and contribute to community cohesion and empowerment in ways that can spread to other development activities. . . .*
>
> — World Bank 1993: 55

By 'collective action optimists' I refer to those social scientists and development practitioners who assume that wherever participation in collective action is required for the mutual benefit of a group of people, it will naturally occur.

At present, trust in participation is remarkably strong among all international development organisations and has become a compulsory ingredient of most projects. Participation optimism originates from orthodox group theories prevailing in political science in the 1950s, when it was generally believed that the existence of a collective interest constitutes a sufficient motive for joint action. Accordingly, if given a chance, people would try to influence decisions that affect their lives. Failures to live up to these expectations were considered abnormalities, signs of individual or systematic pathologies.[2]

In the Western world low participation in elections, voluntary organisations and collective action in general led political scientists to question the validity of these assumptions already during the 1960s. The costs of participation were recognised as a factor that may induce individuals to take a 'free ride' on other people's efforts instead of doing the work themselves. This paradigm shift, however, had rather marginal influence on international development agencies and participation advocates. In spite of the growing volume of empirical evidence showing that attempts to foster participation are often far from successful, participatory development remains an established orthodoxy among development agencies across the whole political spectrum.

The emphasis on participation is particularly strong in the domain of irrigation projects, where farmers' critical role to achieve project objectives has been recognised for several decades. Until the1970s, however, governments and international development agencies, although investing billions in hydraulic infrastructure, hardly paid any attention to the development of local organisations. Although it was understood that irrigated agriculture requires organised collective effort to secure water, it was assumed that water control groups at the tertiary level would automatically evolve because they were needed. Above the field level, government agencies were assumed to be best qualified to manage modern hydraulic infrastructure. By the mid-1970s it became clear that the outcomes of capital intensive investments in water resource development

> . . . differed distressingly from those that had been expected: less land was served than originally planned; cropping patterns were not modified as intended; water users failed to complete local drainage and irrigation facilities. — Coward 1980: 7

---

2. For a critical summary of orthodox group theory see Nagel (1987) and Olson (1965).

During the same period it was also recognised that government agencies of developing nations did not have the financial and administrative capacity to operate and maintain all their infrastructure. Passing the burden of O&M to the people in the name of 'participation' thus became an imperative.

Whereas in other development domains the concept of participation was often associated with empowerment, this is not the case in the domain of irrigation. In this field participation had an instrumental, utilitarian meaning and was considered a necessary means to an end. Thus, farmers' participation primarily referred to the need to mobilise labour and material resources from the users for the operation and maintenance of hydraulic infrastructure. Only more recently the definition of participation or the list of tasks in irrigation management in which farmers could effectively participate was extended to include virtually all activities related to the project cycle, from project identification to evaluation.

In the water sector, participation is generally valued as a way to prevent unsatisfactory performance of hydraulic infrastructure. More "participation" through farmers' organisations is often proposed as a solution even in situations where there is no evidence that lack of participation is the problem. This tendency has led to a persistent conceptual confusion between participation and the formation of so-called 'beneficiary organisations.'

Critical views about the current faith in participation and about what Rahnema (1992: 117) defined as 'manipulated or teleguided forms' of institutionalised participatory approaches began to emerge over the last decade. Their focus generally is on institutional issues, on the political will of the concerned agencies to promote 'real participation' or on the viability and effectiveness of the adopted strategies.[3] In general, however, critiques of participatory approaches to development do not address issues that are so prominent in collective action and in participation theories when applied to Western countries, namely whether rational self-interested people are likely

---

3. See for example Hall 1988 and Rahnema 1992, Ambler 1993, Mayoux 1995.

to invest their time and resources to participate in collective action.[4] Indeed, only limited attention has thus far been paid to the perspectives of those who are expected to 'participate.' Questions about whether or under what circumstances, people whose participation is expected are willing to bear the costs of participation and whether they recognise themselves as an interest group willing to pursue a common goal are rarely raised. This is particularly striking in the domain of irrigation management. The assumption that it is sufficient to persuade people about their collective interest in sustaining a public development investment is one of the causes of the unsatisfactory outcome of so many endeavours to enhance participation. Other factors, such as the lack of motivation of government officials to turn over irrigation systems to users' associations and their 'perverse incentives' to build public infrastructure of poor quality and to allow for its rapid deterioration, are equally important, although they will be only marginally addressed by this book.[5]

Collective Action Pessimism

> *. . . the customary view that groups of individuals with common interests tend to further those common interests appears to have little if any merit.* — Olson 1965: 2

During the same period when in the domain of irrigation development there was growing enthusiasm about farmer's participation and about turning over to system users significant responsibilities in the operation and maintenance of public infrastructure, diametrically opposed views started to emerge in economics, political science and other disciplines. Three distinct paradigms have been particularly influential in supporting theories about the limited opportunities for people to further their common interests: the 'logic of collective action,'

4. In the domain of irrigation management the costs participation and the problem of free riding have been recognised for some time. Freeman (1989) and Parlin and Lusk (1991) recognise the obvious fact the participation in operation and maintenance (O & M) of irrigation systems represents a cost for the individual farmer, which he would rather avoid. However, only Small *et al.* (1989) go as far as arguing that only enforcement through sanctions makes participation to O & M possible.
5. On this subject see for example Repetto (1988), Gisselquist (1991), Ostrom (1992a).

'prisoner's dilemma' and 'the tragedy of the commons.' Although these three paradigms share some fundamental views about the inherent conflict between individual interests and group interest, each of them has had powerful influence in academic and policy circles.

THE LOGIC OF COLLECTIVE ACTION

The Logic of Collective Action is the title of a book written by Olson in 1965, which even today may be considered the touchstone of collective action theory. Olson's theory is often used to demonstrate that rational individuals are unlikely to participate in a group endeavour to pursue a common goal. By 'collective action', Olson refers to group efforts to further common interests. His logic therefore encompasses almost all acts of participation that are aimed at goals shared by a group of people. These goals may relate to a tangible good (e.g., the maintenance of an embankment) or to immaterial benefits (e.g., the victory of an electoral candidate) but they all have in common that if the goal is achieved, everybody benefits from it regardless of whether he or she contributed to its provision. Economists, including Olson, refer to these sorts of group goals — characterised by jointness of supply and impossibility of exclusion — as 'public goods.' The problems related to the non-excludable nature of public goods and that economically rational individuals would not voluntarily contribute to pay for them was well understood in economics already before Olson. This author, however, recognised the link between collective action and public goods and that all group goals and group interests are subject to the same dilemma. Olson realised the relevance of the theory of public goods for the analysis of collective action, thus exporting its inherent logic to other social sciences. Although he does not deny the possibilities of collective action, his theory provides perhaps the most articulated and sharpest contrast to those who take the formation of interest groups for granted. Such theories, according to Olson, are 'logically inconsistent':

> . . . it is *not* in fact true that the idea that groups will act in their self-interest follows logically from the premise of rational and self-interested behavior. . . . The widespread view, common throughout the social sciences, that groups tend to further their interests is . . . unjustified, at least when it is

> based, as it usually is, on the (sometimes implicit) assumption that groups act on their self-interest because individuals do.
>
> — Olson 1965:1-2

Unlike Hardin's *Tragedy of the Commons* and to the *Prisoners Dilemma*, Olson does not deny the possibility of rational individuals' pursuing a public interest; rather, he offers a radically different account of the logical basis of organised collective action. As will be discussed later in this chapter, size and other characteristics of a group are considered by Olson of central importance in determining an individual's participation in group action. In his view there may well be some voluntary action in support of a common purpose in small groups but participation in large groups, if it occurs, needs a different explanation.

Coercion by government is considered the obvious source of many collective goods. For Olson, however, coercion is only one instance of a broader phenomenon he calls *selective incentive*. Selective incentives are material or social rewards specifically oriented towards those who contribute to a collective action. Collective action, according to Olson, is always accompanied by private incentives to reward contributors or to punish non-contributors. Selective incentives are not oriented towards the group as a whole like a collective good but operate selectively, towards the individuals in the group. They must be selective, so that those who do not join the organisation and work for the group interest can be treated differently from those who do. Only selective incentives or private benefits would stimulate a rational individual to act in a group-oriented way. Selective incentives are one of Olson's central themes and may be considered his 'simple and sovereign' theory of collective action (Marwell and Oliver 1993: 5).

Olson's logical theory should be known to those development theorists and practitioners who are excessively optimistic about solving many of the development constraints through the formation of groups. However, it should not be embraced indiscriminately either. As argued by Marwell and Oliver:

> Free riding is a real problem. And yet collective goods are everywhere provided. [...] Any reasonable theory must account

> for these phenomena as well as for the equally obvious fact that many collective goods ardently desired by some groups or even a whole population, never come to pass.
>
> — Marwell and Oliver1993:6

There are many factors that may explain collective action other than those discussed by Olson. First of all, as pointed out by Melucci (1995: 18), it is necessary to overcome 'the Olsonian individualism;' 'the naïve assumption that collective phenomena are simply empirical aggregations of people acting together.' Secondly, by also considering non-material rewards as acceptable selective incentives and by recognising that also 'extra-rational motivations' (such as moral motivations and self-realisation) may determine individuals' participation in collective action, it is possible to recognise many more situations under which it may occur. For example, the participation in women's movements or environmental groups may be a highly gratifying social and moral experience that has little in common with the burden of cleaning an irrigation canal.

Finally, one cannot understand the emergence of collective action, or its failure to emerge, by focusing on group characteristics and selective incentives, without keeping in consideration the purpose of collective action.[6] The fact that all collective actions have in common that individuals cannot be excluded from benefiting from their achievement does not mean that fighting, for example, for the abolition of genetically manipulated reproduction technology in Western Europe follows the same logic as contributing to raise an embankment for flood protection in rural Bangladesh.

Although group attributes — as will be further discussed later in this chapter — may be important to understand the dynamics and outcomes of collective action, the broader cultural, political, socio-economic and ecological context needs to be kept in close consideration as well.

---

6. Olson does not give much importance to the specific purpose pursued by an organisation because — he argues — there is some question whether there is any single purpose that would be characteristic of organisations generally (Olson, 1965:5).

PRISONERS' DILEMMA

The *Prisoners' Dilemma* originates from mathematical game theory, which was one of the dominant frameworks for analysing social interactions in the 1950s and 1960s. It was developed by Flood and Dresher, who were concerned with testing solutions for non-cooperative games. It was named *Prisoners' Dilemma* only later, by A.W. Tucker, a game theorist of Princeton University (Hardin 1982). Judging from the amount of publication on the *Prisoners' Dilemma*,[7] this is perhaps the most well known and appealing game. It shares with Olson's theory of collective action, its generality and its apparent power in providing a solid basis for a profoundly disturbing conclusion — that rational people cannot achieve rational collective outcomes. In social sciences the *Prisoners' Dilemma* is less known and hence its underlying parable needs to be summarised briefly here:[8]

> Two suspects are being separately interrogated about a crime they jointly committed. They know that if they both stay silent they will receive a light prison sentence. If one stays silent while the other confesses the first will receive a long prison sentence while the other goes free. If both confess they both receive a medium prison sentence. Each person can choose only once — which means that if one chooses to stay silent while the second confesses the first cannot then confess upon learning of his sentence. This is what creates the dilemma. Their joint interest is for both not to confess (that is, for them to 'cooperate' with each other). But the outcome is that both confess (both 'defect'). From the point of view of either one of them, staying silent while the other confesses would give the worst outcome, and confessing at least ensures that this outcome is avoided while it also opens the possibility that the confessor will go free if the other stays silent.
>
> — Wade 1988: 201

---

7. In 1975 Grofman and Pool found that more than 2,000 articles had been devoted to Prisoners' Dilemma (Hardin, 1982).
8. I choose here Wade's (1988) description of the parable for its somewhat more narrative and less abstract style than the one provided by authors accademically closer to game theory (e.g. Hardin 1988; Runge 1992, Taylor, 1987).

The *Prisoners' Dilemma* suggests in a clear manner that it is impossible for rational people to cooperate, a conclusion that bears directly on fundamental issues in ethics and political philosophy. Indeed,

> the paradox that individually rational strategies lead to collectively irrational outcomes seems to challenge a fundamental faith that rational human beings can achieve rational results. — Ostrom 1990: 5

Like Olson's logic of collective action, the *Prisoners' Dilemma* has been applied to a broad spectrum of situations. Several theorists have relied on this argument to provide the essentials of a theory of state, which would be needed above all to enforce contracts and punish deviants, so that social order can be maintained.

Its application to explain the tragic fate of common resources has been criticised by many scholars.[9] In fact, not only does empirical evidence prove differently but the way the game is structured has generally little in common with real life situations. Runge (1992), for example, calls attention to the fact that the game represents a special case of joint action that can only be understood if one recognises the structure of the game as a function of the institutional environment in which it is embedded. Those who see in the *Prisoners' Dilemma* an inevitable human behaviour tend to confuse cause and effect and that the result of the game is just an artefact of the way in which it is set up. If the rules for exacting confessions from apprehended prisoners are different, they will also have different strategies and better reasons to cooperate. Thus, as argued by Bromley:

> It is essential to understand that the institutional structure of any game (or life situation) reflects the prior social purpose to be served by the human interaction under consideration. . . . The existing institutional structure reflects, among other things, prevailing cultural and social norms regarding individualism and its relation to collective notions. In that sense we can say that people behave (or choose) in an institutional context — not a very surprising observation really. — 1992: 6.

9. See Bromley (1992), McCay and Acheson (1987) and Ostrom (1990) for comprehensive and empirically grounded critiques to the application of the Prisoners' Dilemma to common property resource management.

The applicability of the *Prisoners' Dilemma* to real life collective action problems is also questioned by Wade (1988) for whom two key assumptions must hold if a situation is to be plausibly modelled as a *Prisoners' Dilemma* and its pessimistic conclusions drawn. The first is that the players choose in ignorance of each other's choices, that they have no contact with each other, cannot negotiate themselves and change the rules and have no previously established shared values or moral codes of behaviour. The second is that each player chooses only once before the payoffs are received and so cannot change his mind upon finding out what the other has done. Quite obviously, where the situation is an enduring and recurrent one, the logic changes and individuals may find it convenient to cooperate. In particular, if they know what others have chosen and can alter their choice accordingly, the rational strategy may be one of 'conditional cooperation' or 'cooperate first, defect if the others defect' (Wade 1992: 203).

THE TRAGEDY OF THE COMMONS

> *Ruin is the destination toward which all men rush, each pursuing his own best interest in a society that believes in the freedom of the commons. Freedom of the commons brings ruin to all.*
>
> — Hardin 1968: 1244

The still widely used metaphor 'tragedy of the commons' owes its origins to an article by Garrett Hardin appeared in *Science* in 1968. To make his point about the inevitable conflict between individual behaviour and collective interest, Hardin invites the reader to picture a pasture open to all. In such a situation a rational herdsman will seek to maximise his gain by adding more and more animals to his herd. The tragedy is caused by the fact that each herdsman will act the same way as 'each being is locked into a system that compels him to increase his herd without limit in a world that is limited' (Hardin 1968: 1244).

The powerful impact of this article may partly be explained by the time and socio-cultural context in which it was published. It was in those years that the Western world became aware of the dramatic consequences of an unconcerned use of natural resources that was rapidly leading to their depletion and to an irreversible loss of biodiversity. The same years were also characterised by pro-state intervention ideologies and it is not surprising that Hardin's thesis was used to support policy prescriptions in favour of bringing common property resources under state control. This indeed became the faith of many communal resources in South Asia, where government has long been equated with centralised state bureaucracy and the potential for various local authority systems has consistently been overlooked. For those people whose livelihood depended on these resources, as well as for the resources themselves, the consequences have often been tragic, as cases from Nepal, India and Bangladesh demonstrate.[10]

The paradigm continued to exercise a strong influence a decade later, when trust in public policy was replaced by a new faith in privatisation. The causes of environmental degradation and economic loss are assumed to be in the nature of property rights, while the role of more complex features of the socio-economic systems in which these tragedies occur are being ignored. As pointed out by Bromley and Cernea (1989), considering that many private natural resources are also subject to unsustainable uses, it is difficult to follow the conceptual logic of privatisation recommendations.

Besides its influence on policy circles, Hardin's theory has revived the interest in common pool resource management in anthropology and other social sciences. Since the publication of Hardin's article, several studies were carried out or reinterpreted to prove that the 'tragedy of the commons' is not the universal rule. Indeed, world history provides many concrete examples that prove that human societies have the capacity to develop institutions and enforce

10. The dramatic consequences of bringing common pool resources under state control are well documented in the case of India and Nepal (Jodha 1986; Bromley and Chapagain 1990). The issue has received less attention in Bangladesh, where common pool resources (in particular surface water bodies) have been brought under central government control and hence leased out to private individuals with the aim to 'finance rural development from below' (Blair 1989). Nevertheless there is enough evidence that also in Bangladesh such a policy favors the rich and influential, deprives the poor of their main source of animal proteins, and contributes to the depletion of fish resources (Duyne 1999b).

restrictions to prevent the depletion of natural resources.[11] This rich collection of cases shows that the premises of Hardin's parable, according to which each individual acts essentially alone as if there were no institutions to condition his behaviour and as if he had no understanding of the consequences of his use of resources, are wrong. It may thus be concluded that 'Hardin's paradigm is not only socially and culturally simplistic, it is historically false' (Bromley and Cernea 1989: 7).

Group Attributes

The two preceding sections presented two different perspectives about the possibility of collective action. The first is characterised by an overly simplistic optimism about people's capacity to cooperate and the second by sweeping pessimism. These extreme positions have triggered a lively debate and systematic research that has contributed to significant advances in understanding the conditions under which collective action emerges.

The point of departure for most research on determinants for collective action is Olson (1965), who explains collective action by focusing on group attributes. Whereas the emphasis on group attributes is particularly strong among orthodox collective action theorists, the importance of related variables is also recognised by scholars concerned with the conditions required for the emergence of organisations capable of ensuring a sustainable management of common pool resources.[12] In this section I shall discuss the importance of group attributes by focusing on five key issues, namely group size, heterogeneity, interdependence, the role of organisations and the importance of 'community'.

GROUP SIZE

> *Two neighbours may agree to drain a meadow, which they possess in common; because it is very easy for them to know each others mind; and each must perceive, that the immediate*

11. See for example Netting (1981); McCay and Acheson (1987), Wade (1988), Ostrom (1990).
12. See for example Bromley *et al.* (1992), Bromley and Cernea (1984), Oakerson (1992) and Ostrom (1990; 1992a; 1992b).

> *consequence of his failing in his part, is, the abandoning of the whole project. But 'tis very difficult and indeed impossible that a thousand persons shul'd agree in any such action; it being difficult for them to concert so complicated a design and still more difficult for them to execute it; while each seeks a pretext to free himself of the trouble and expense and would lay the whole burden on others.* — David Hume, 1878[13]

One of the most controversial issues in contemporary literature on collective action is related to the effect of group size on the likelihood of group action. According to Olson (1965) a clear distinction ought to be made between the behaviour of individuals depending on some basic characteristics of the group to which they belong. To explain this point he uses two parallel typologies. The first one is based on group size whereby a distinction is made between small, intermediate and large groups. The second typology differentiates between 'privileged' groups and 'latent groups.' A privileged group is defined as a group in which each of its members or at least some of them have an incentive to see that the collective good is provided, even if he has to bear the full burden of providing it himself. In such a group the collective good may be obtained even without any group organisation or coordination (Olson 1965: 50). The concept of 'privileged groups' relates to the assumption of groups being heterogeneous and also draws attention to the role of organisations, two important issues that will be discussed further below. A latent group is defined as a very large group distinguished by the fact that whether a specific member contributes or not to provide the collective good, will not significantly affect other members. Therefore none has any reason to act. The distinction between latent and large groups is less clear and the two terms are almost used interchangeably.

According to Olson, small groups may be able to provide themselves with a collective good simply because of the attraction the good may have for the individual group member. The greater the interest of any single group member, the greater the likelihood that the member will get such a significant portion of the total benefit from the collective good that he will gain from seeing that the good is

13. David Hume, *A Treatise of Human Nature*, cit. in Hardin (1982: 40).

provided, even if he has to pay all the cost himself. The distribution of the burden of providing a public good in a small group is never proportional to the benefits conferred by the collective good. Olson even argues that in small groups "*. . . there is a systematic tendency for 'exploitation' of the great by the small !*"[14] Selective incentives are not required in small groups; they may have the capacity to provide themselves with a collective good without relying on coercion or any positive inducement apart from the collective good itself. This is because each of the members or at least one of them, will find that his personal gain from having the collective good exceeds the cost of providing some amount of that collective good (Olson 1965: 34). In small groups free riding is not really an issue because each member or at least some among them, is sufficiently interested in the collective good to ensure its provision. The risk of free riding is further checked by the fact that an individual's shirking behaviour would be noticeable and sanctioned.

In 'intermediate groups' no single member gets a share of the benefit sufficient to give him an incentive to provide the good himself. However, intermediate groups do not have so many members that no one member would notice whether any other is or is not helping to provide the collective good.

For Olson it is not enough to know whether a small group will provide itself with a collective good; it is also necessary to determine whether the amount of the collective good that the group will obtain is optimal for the group as a whole. The unequal interest among the members of a group in the collective good would explain why even in small groups the provision will be strikingly sub-optimal. In fact, once members with the largest fraction of the public good have obtained the amount they want, no one has an incentive to provide more of the collective good.

---

14. Olson (1965: 29). Italics by the author, whereby he underlines in a footnote that "the moral overtones of the world 'exploitation' are unfortunate; no moral conclusion can follow from a purely logical analysis." The word exploitation is chosen because it is commonly used to describe where there is a disproportion between the benefits and the sacrifices of different people.

The likelihood of large groups being capable to provide themselves with a collective good is considered by Olson to be very small. The larger a group is, the farther it will fall short of obtaining an optimal supply of any collective good and the less it is likely to obtain even a minimal amount of such a good. In short, the larger a group, the less it will further a common interest (Olson 1965: 36).

Group size is considered an important variable beyond orthodox collective action theory as well. Ostrom (1992), in the context of her comparative analysis of a number of case studies on common pool resource management, concludes that effective organisations are more likely to emerge where the number of 'appropriators' is sufficiently small to keep the cost of communication and decision-making relatively low.[15] Nevertheless, beneath an apparent consensus, the effect of group size on collective action remains a controversial issue, with more recent studies emphasising that small size groups may not be considered a general precondition for a stronger ability to perform collectively.[16]

HETEROGENEITY

The above summary of Olson's theory of how the size of the group determines collective action calls attention to an equally important variable, namely *heterogeneity.* In fact, Olson's definition of 'privileged group' is closely related to the assumption of groups being heterogeneous. His concept of heterogeneity, however, does not relate to socio-economic or cultural differences among group members, but to the differential value individuals place on the public good and to the resources available to contribute to its provision. The fact that generally not all individuals belonging to a group share a similar interest in a public good may appear quite obvious, but it constitutes one of Olson's more important contributions and has had a strong influence on further collective action research. Although homogeneous groups

---

15. Ostrom's concept of appropriation is closely linked to her definition of common pool resources, whereby she calls the process of withdrawing resource units from a resource system 'appropriation.' Appropriators are thus defined as those people who withdraw such units and may refer to herders, fishers, irrigators, etc. Ostrom (1990: 30).

16. See for example Vedeld (2000).

may exist, it is indeed misleading to treat heterogeneous groups as if they were homogeneous and to examine only the aggregate group interest in a collective good. While for Olson heterogeneity explains why some groups are more likely to succeed in collective action than others, this factor is also considered the main reason for a sub-optimal provision of the collective good. The influence of group heterogeneity on collective action has attracted many scholars and has been subject to sophisticated mathematical modeling.[17]

According to Marwell and Oliver (1988;1993) collective action usually depends on a relatively small cadre of highly interested and resourceful individuals rather than on the efforts of the 'average' group member. These individuals behave differently from a typical group member, for example, by assuming leadership positions, covering initial costs and making up for group members who fail to contribute their share.

However, heterogeneity does not necessarily increase the probability of collective action. Hardin (1982) analyses the issue of heterogeneity in terms of 'asymmetries' that complicate the analysis of collective action. Inequality or 'asymmetries in demand for a collective good' may lead to polar opposites: to enhance or dissipate prospects for collective action. Dissipation is likely when the collective good sought by a group may also be obtained privately, although perhaps at a higher cost. In such a situation the intense demanders may opt for the private good even if it has some disadvantages.

Over the last decade heterogeneity and its influence on collective action has gained attention of a number of researchers.[18] These scholars, however, do not rely on Olson's definition of heterogeneity, as they are more concerned with the effect of social, economic and cultural heterogeneity on collective action. They emphasise the existence of different types of heterogeneity and recognise that the degree of heterogeneity among members of a group may vary with respect to a number of variables, such as political influence, wealth, endowments, economic interests, culture, exit opportunities, etc.

17. Cf. Heckathorn (1994) and Marwell, *et al.* (1988).

18. See, for example, Schlager and Blomquist (1998), Vendeld (2000), Bardhan and Dayton-Johnson (2000) and Bosc and Weber (2000).

Ostrom, for example, considers heterogeneity an important variable that may help explain the emergence of what she calls 'appropriator organisations.' The author argues that homogeneity among appropriators, defined in terms of the group not being strongly divided — by natural boundaries, conflictive resource use patterns, different perceptions and substantial exposure to risks and cultural antagonisms — may increase the capacity of a group to organise for an effective management of common pool resources (Ostrom 1992: 299). This assumption is based on a review of numerous case studies of resource users failing to address shared problems; all cases point to some sort of heterogeneity as one of the principal causes for failure.

These pessimistic views about a heterogeneous group of people's likeliness to cooperate are not supported by Bardhan and Johnson (2000). Based on case studies from India and Mexico, they argue that collective action in unequal societies is not necessarily bound to fail, among other reasons, because:

> Highly unequal agrarian societies might sometimes exhibit widespread adherence to a hierarchical ideology that facilitates monitoring and enforcement of cooperative agreements. — Bardhan and Johnson 2000: 5

An interesting contribution to the analysis of the role of heterogeneity in collective actions stems from Vedeld (2000). Based on research among the Fulani in Mali, Vedeld reaches the conclusions that there is little direct relationship between the degree of heterogeneity and the success in collective action, except when heterogeneity among leadership and elite groups is considered. Vedeld (2000: 15-16) thus found that:

(a) Homogeneity among elite groups enhances the capacity for collective action;

(b) When heterogeneity in economic interests between elite groups intensifies and coincides with other dimensions of heterogeneity, collective action becomes difficult to achieve;

(c) Collective action is enhanced by political elites and leaders being a bit better off than the average community member, but not when their assets were mobilised against the economic interests and fair treatment of other social groups;

(d) Coordination power of the leadership related to the management of common property resources is undermined when leadership has extensive recourse to state officials external to the village community.

The current controversy about the role of heterogeneity in collective action is to some extent related to the different definitions given to this concept. Olson's original definition of heterogeneity, simply focusing on the differential value individuals belonging to a group may place on a public good, is still useful in explaining why other types of group heterogeneities and in particular socio-economic inequalities are not necessarily an obstacle for collective action.

INTERDEPENDENCE

> *I shall argue that corporate organisation. . . is likely only when external costs are high — when, in other words, the interdependencies in production are such that any one cultivator is exposed to a high risk of crop loss and social conflict as a result of the activities of other people.*
>
> — Wade 1988: 14

Early collective action theorists, including Olson, built their theorems upon the assumption that individuals make isolated, independent decisions whether or not to contribute to collective action. Nowadays, the critical role of interdependence between group members is recognised by formal collective action theorists as well as by scholars concerned with common pool resource management.

Marwell *et al.* (1988) were among the first collective action theorists who challenged the assumption that individuals decide independently. They contended that in most decision-making situations, people are aware of what others are doing, as they often have social relations that make influence or even sanctions, possible. Independent decision-making may apply to very large or 'latent' groups but not to the more frequent, smaller groups. It may thus be assumed that people take into account whether and how much others have already contributed and that their decisions follow a sequential pattern.

Marwell *et al.* (1988) argue that the organisation of potential contributors, making their decisions interdependent, is a requirement to overcome the free-rider problem. However, they do not consider the potential to organise a group to be necessarily always present. The possibility of group organisation depends on the social ties in the group, particularly the density and frequency of ties on the extent to which they are centralised in a few individuals and on the cost of communicating and coordinating actions through these ties. The overall density of social ties in a group improves its prospects for collective action whereby it is noticed that network centralisation is beneficial for collective action.

Within the study of common pool resources, the concept of interdependence refers to a distinct problem, namely that,

> . . . when multiple appropriators are dependent on a given CPR as a source of economic activity, they are jointly affected by almost everything they do. Each individual must take into account the choices of others when assessing personal choices. — Ostrom 1990: 38

Ostrom (1990) notices that if appropriators disregard their interdependence in relation to a common pool resource by acting individually, the total net benefits they will obtain usually will be less than could have been achieved if they had coordinated their strategy in some way. Thus, according to Ostrom (1990), it is interdependence that generates a need to organise.

Wade (1988) considers the high interdependence in agricultural production prevailing in peasant societies to be crucial for understanding the emergence of collective action. He notices that farmers in peasant societies are interdependent to the extent that what one farmer does has repercussions for others. Yet they generally decide independently. Collective action through village-based corporate organisations would thus represent 'an adaptation to the disjunction between interdependence in production and the private decision-making system which directs agriculture, by which the scope of private decision-making is reduced and the scope of collective decision-making increased'(Wade 1988:14).

As may be noticed, although each of the above-mentioned authors gives a different meaning to the concept of interdependence, they all link it with the subject of organisation.

ROLE OF ORGANISATION

The participation concepts and strategies embraced by international development agencies attribute a critical role to the formation of organisations, without explaining the link between group action and organisation and without much concern for the conditions under which organisations are necessary, likely to emerge and capable of surviving.

In the domain of irrigation management, the importance of establishing organisations to enhance participation is stressed, for example, by Uphoff (1986a), who argues that farmer organisations are required to make participation more 'effective'. The lack of formal organisations is considered by Cernea (1987) as the principal cause of rural underdevelopment and for the failure of many projects:

> All over the world, the degree of formal organisations of rural societies lags far behind that of urban populations. This is a fundamental characteristic of rural underdevelopment that accounts largely for the weakness and vulnerability of rural societies. Many rural programmes collapse precisely for want of grass-root organisations able to sustain group action. . . . Farmers' organisations, pastoral associations, cooperatives, credit groups and water users' organisations, for example, are vital engines for development. The numerous informal organisations in traditional societies can often be a matrix for structuring stronger formal organisations.
>
> — Cernea 1987: 11

In development practice participation and organisations are almost treated as synonyms. This conceptual confusion has led to acceptance of the number of organisations formed in the context of a development project as a valid indicator of the degree of 'participation'.

These days, however, there is an increasing awareness that the enthusiasm about establishing new organisations to enhance participation in development projects has all too often led to the formation of organisations that only exist on paper.[19]

The assumption that the formation of organisations will automatically trigger collective action dismisses insights gained by collective action theorists on the relation between participation, collective action and organisation. In fact, the role of organisations in collective action is already extensively discussed by Olson, who argues that:

> Organisations . . . perform a function when there are common or group interests and though organisations often also serve purely personal, individual interests, their characteristic and primary function is to advance the common interests of groups of individuals. —Olson 1965: 7

Olson specifies that the need for organisations to pursue collective action depends on the size of the group. He does not consider the formation of organisations necessary for small groups, in which by definition free riding and shirking would be noticeable. Moreover, given the fact that groups tend to be heterogeneous, in small groups it is likely that an individual member with a high interest in the good is willing to contribute to its provision regardless of whether other group members do the same. Thus, in small groups it may not be worthwhile to bear the high transaction costs involved in establishing an organisation, an important conclusion that should be kept into account by external development agencies promoting the formation of organisations.

The situation changes in intermediate groups. Olson argues that for this type of group, a collective good may or may not be obtained but 'no collective good will ever be obtained without some group coordination or organisation' (1965: 50). This is not the case in 'privileged groups', in which the good, like in small groups, may be obtained without any group organisation or coordination.

---

19. Ambler (1993); Barenstein (1994a); Duyne (1994a) and (1994b).

Organisations also are given a critical role in collective action studies focusing on social movements. According to Melucci (1995), for example, social movements survive if they can resist centrifugal forces, that is, if they can develop relatively stable organisations and leadership. However, this leads to institutionalisation and bureaucratisation, which may transform the objective of the collective action. In Bangladesh, the rather negative experience many people have made with corrupt formal organisations and the tendency of rural elites to capture local organisations for their personal interests may explain why people with a common interest in improved water management are not necessarily inclined towards establishing permanent organisations.

In any case, the relation between collective action and organisations is far more complex than was formerly assumed. This is recognised by Ostrom (1990), who makes a distinction between 'being organised' and 'forming an organisation':

> . . . the problem facing CPR appropriators is one of organising; how to change the situation from one in which appropriators act independently to one in which they adopt coordinated strategies to obtain higher joint benefits or reduce their joint harm. *That does not necessarily mean creating an organisation*. Organising is a process; an organisation is the result of that process.
>
> — Ostrom 1990:39; emphasis added

Ostrom underlines the importance of establishing appropriators' organisations, although she considers them a necessary but not sufficient condition for avoiding the tragedy of the commons. Ostrom is aware that organisations do not always emerge where they are needed and that the conditions for the emergence of such organisations need to be examined. Based on her review of numerous case studies, Ostrom (1992) notices that the emergence of appropriators' organisations is associated with some prior experience with organisations such as: (a) general purpose organisational structures (such as village councils or cooperative organisations); (b) specialised organisational structures related to the resource without prior management responsibilities and (c) nearby organisations that have helped others to solve similar CPR management problems.

Wade's study on collective action in south India clearly shows the importance to distinguish between collective action, organisations and institutions.[20] Wade (1988; 1992) links the concept of collective action to the emergence of a 'public realm' consisting of four main 'corporate institutions' consisting in his case of informal village councils, village standing funds (mobilised in various ways by the village councils), a work group of village field guards and a work group of common irrigators. The concept of corporate institutions as defined by Wade is particularly relevant to this study. In fact, as will be discussed later, also in Bangladesh collective actions are rarely associated with the establishment of specific organisations. One of the four regions studied, however, is characterised by the presence of 'corporate institutions' astonishingly similar to those found by Wade in some villages in south India.

Finally, an entirely different explanation of why formal organisations are often absent and not required in relation to a particular group action is provided by Olson's 'by-product theory'. Olson (1965) uses this concept to explain the political power of some large groups. Hence, the membership and power of large pressure-groups would not derive from their lobbying achievements but should be considered a by-product of their other activities. This is considered a necessary explanation for the existence — against all odds — of large voluntary organisations. The by-product theory has the merit to explain that some groups may not need to be organised because they are already organised for other purposes. In rural Bangladesh, the obvious social organisation upon which groups of people may rely for collective action is the village. This calls attention on a critical

20. In development practice the terms 'institution' and 'organisation' are often confused, whereby the expression 'institutional development' is often used to refer to the process of establishing organisations. However it is crucially important to make a distinction between these two terms. Institutions may be defined as 'the rules of the game in society or, more formally, . . . the humanly devised constraints that shape human interaction' (North 1990:3), whereas organizations are to be understood as 'groups of individuals bound by some common purpose to achieve objectives.' (North 1990:5).

variable that has recently been 'discovered' by collective action theorists, namely, the role of 'community'.

THE ROLE OF 'COMMUNITY'

The importance of community for collective action was emphasised by several authors to explain why some users of common resources are able to solve their collective action problems endogenously, while others are not.[21]

Singleton and Taylor (1992: 315) define community as a set of people: (i) with some shared beliefs, including normative beliefs and preferences beyond those constituting their collective action problem; (ii) with a more-or-less stable set of members; (iii) who expect to continue interacting with one another for some time to come; and (iv) whose relations are direct (unmediated by other parties) and multiplex. They maintain that a community group is more likely to arrive at endogenous solutions to its collective action problem if it comprises individuals who are mutually vulnerable. If a group has these attributes it is more likely to be capable of facing the transaction costs related to collective action, which are considered the main reason why many collective action problems remain unsolved. The authors, giving much importance to whether a group is able to find an *endogenous solution* to a collective action problem, argue that:

> . . . when there is a great deal of community, the group is likely to solve its collective action problem not merely endogenously but also without any or with only minimal specialised institutional roles.
>
> — Singleton and Taylor 1992: 316

Singleton and Taylor's (1992) theory, however, is not very persuasive. Community as defined by Singleton and Taylor (1992) may be a *necessary* but certainly not *sufficient* condition for the endogenous solution of common problems (Ostrom 1992). It remains, however, a misleading term which tends to overplay harmony and cohesion and downplay social and economic inequalities and power differentials

21. See Taylor (1987), Singleton and Taylor (1992), Taylor and Singleton (1993), Ostrom (1992) and Mearns (1995).

within a group. While the characteristics that, according to Taylor and Singleton, define a 'community' may be considered relevant variables in the context of a comparative analysis of collective actions, it may be better to de-link them from a term whose definition remains controversial and whose connotation, especially in the context of development, is far from being neutral.

The Role of the Context

Collective action theory, with its focus on group attributes has generally not paid much attention to the influence of contextual factors on people's need and willingness to cooperate. But lack of knowledge about contextual factors may lead to simplified judgements about why some groups are more successful in overcoming collective action problems than others.

The conditions under which collective action is likely to occur have recently attracted an increasing number of researchers concerned with common pool resource management.[22] Some explicitly build upon collective action theory but also recognise that there are contextual factors that need to be appraised to understand why certain groups are more successful than others in overcoming collective action problems. Contextual information is often given only as a background, without being integrated in the analysis. This section examines a set of contextual issues that may affect the capacity to engage in collective action. Emphasis will be given to three specific domains: the ecological conditions, the socio-economic conditions and the role of the state.

ECOLOGICAL CONDITIONS

Cultural ecology has made substantive contributions to understanding the relation between the environment and patterns of social organisation.[23] So-called 'hydraulic societies' were already pointed

22. See for example Bromley (1992), McCay and Acheson (1987), Oakerson (1992); Ostrom (1990), (1992a), (1992b); Wade (1992), Edwards and Stein (2000).

23. See for example Leach (1957), Downing and Gibson ed. (1974), Netting (1981) Hanks (1972), Geertz (1972). Though these studies were not produced with the aim of providing explanations for sustainable patterns of natural resource management, they provide the empirical foundations to the theoretical models developed by authors such as Ostrom (1990, 1992) and Bromley (1992).

out by Marx as a distinct societal type.[24] The link between hydraulic infrastructure, social complexity, bureaucracy and ultimately a tendency towards autocratic centralised political systems has been dramatically illustrated by Wittfogel (1957), whose work triggered valuable research of numerous irrigation societies.[25]

From a different perspective, the importance of ecological variables to explaining collective action has been underlined by Oakerson (1992) and by Ostrom (1990; 1992b). Their focus, however, is primarily on the specific characteristics of the common resource system under consideration and less on the ecological context of collective action in general.

An important contribution to understanding the relation between the natural environment and the emergence of collective action is offered by Wade (1988). After years of research on various social and institutional aspects of irrigation organisation.[26] Wade reaches the conclusion that collective action is not so much explained by the organisational complexity of irrigation but by agro-ecological factors. Wade found that several villages in south India had developed wholly locally and autonomously a set of 'corporate institutions' consisting of 'village councils', 'village standing funds', 'field guards' and 'common irrigators'. A detailed description of the functions and organisation of these institutions shows how village councils have the authority to regulate matters such as grazing, plantation dates and irrigation as well as to take decisions affecting the whole village. They also administer the village funds, which are ingenuously mobilised, for

---

24. Although Marx and Wittfogel included in their definition of hydraulic society those whose production system relied on large public works such as flood control and drainage systems (such as for example China), this type of infrastructure and its impact on society has not received much attention later on. This factor — and the fact that aridity is considered one of the key ecological attributes of hydraulic societies — explains perhaps why Bangladesh has never been recognized as a 'hydraulic society' and failed to attract researchers with an interest on this subject. Thus, up to date there has been a remarkable paucity of anthropological research focusing on social adjustments to the environment in Bangladesh.
25. See for example Hunt and Hunt (1976), Leach (1959), Geertz (1972) and Lansing (1983).
26. See for example Wade (1979) and Wade (1982).

example, by imposing fines on the owners of animals grazing in restricted areas by 'licensing' the illegal distillation and sale of liquor, and by 'auctioning' or 'franchising' the access to grazing areas for sheep and goats. These funds are used to pay the salaries of field guards employed by the village council to protect the crops against depredation by livestock and thieves and those of the common irrigators, who are responsible to distribute water among the village's irrigated fields.

What caught the attention of Wade was not only that Indian villages were far better organised than what was generally assumed, but also that only few miles would separate villages with a high degree of village-based collective action from villages with virtually none. He noticed that some of the 'dry' villages were better organised than some of the irrigated villages, which thus excluded irrigation as the determinant for collective organisation.

Wade provides one of the rare studies that compare variations of collective actions between culturally homogeneous communities. In fact, the conclusions drawn from his in-depth village study are systematically compared to those of a sample of 41 villages 'located in an area small enough for technology, tastes and general social norms to be constant, while resources such as soil and water vary' (Wade 1988:134). The results show "a clear tendency for corporate organisation to be concentrated in villages of 'tail-end' locations" (Wade 1988:139). These are characterised by a scarce and unreliable supply of irrigation water, poor soils and by sorghum being a more important crop than paddy. These conditions lead to a better scope for stubble grazing and to a high demand of goat and sheep manure. Wade shows how the emergence of corporate institutions is linked to two key ecological factors typical of these villages, namely the *risk* of conflict associated with open-field systems and — rather than the intrinsic nature of water or irrigation — water *scarcity*. Risks are higher and represent a collective problem when a group of people depends on the same and rather scarce resources. Wade thus persuasively demonstrates the crucial importance of local agro-ecological variables in determining the emergence of corporate institutions.

SOCIO-ECONOMIC CONDITIONS

The socio-economic context refers to a wide range of variables related to the organisation of agricultural production, the land tenure system, off-farm employment, access to markets and migration as well as the ethnic and religious composition of the population.

Many participation advocates assume that an unequal distribution of power and productive resources obstructs participation. Esman and Uphoff (1977) make a systematic attempt to compare the performance of over one hundred local organisations, which they consider as being fundamental for development. The authors come to the conclusion that local organisation appears to be more viable where there is relative equity in the ownership and access to productive assets. As opposed to Wade (1988), they find that environmental factors are 'neither often nor strongly associated with performance' (Esman and Uphoff 1977: 26). Their argument relies on case studies carried out by different authors for different purposes with different methodologies. They thus lump together local organisations pursuing entirely different goals, such as women's organisations in Liberia, credit cooperatives in Kenya, irrigation societies in the Philippines and so on. This, together with the fact that they do not make a distinction between endogenous organisations and organisations established in the context of development projects, does not help to make their conclusions very persuasive.

Esman and Uphoff are not the only social scientists who explain the likeliness of collective action by focusing on socio-economic variables. Taylor and Singleton (1993), for example, maintain that a more diversified economic structure and opportunities to migrate make people less interdependent and thus diminishes their need to cooperate, whereas ethnic and religious divisions tend to affect the 'degree of community'. With specific reference to Bangladesh, several authors, based on a recognition that cooperative movements failed to trigger rural development and on other evidence that people have difficulty cooperating, reach the conclusion that the country's socio-economic structure constitutes a major obstacle for collective action.[27]

---

27. See for example Boyce (1987), Jahangir (1982), de Vylder (1982).

While these arguments appear convincing, there is a paucity of empirical research evidence demonstrating how socio-economic factors affect participation in collective actions.

A thorough analysis of the relation between socio-economic inequality and collective action is provided by Wade (1988), who contends that socio-economic variables are far less important to explaining collective action than social scientists generally assume. In fact, Wade's study indicates that endogenous collective organisation can occur in villages characterised by a high degree of inequality. It also shows that within areas that are homogeneous from a socio-economic point of view, there are strong variations with regard to the presence of corporate institutions.

The regions presented in this book have much in common with Wade's villages. They are all characterised by an unequal distribution of power and productive resources, which however, particularly in one region, do not appear to hinder the emergence of collective action. These counterintuitive observation requires careful examination. An analysis of the socio-economic context thus is considered essential to understanding the conditions for collective action.

THE ROLE OF THE STATE

Development sociologists and international aid agencies tend to attribute to the state a central role in triggering collective action through the formation of 'participatory organisations' at the local level. It is generally believed that governmental and non-governmental organisations, with the help of international development organisations, should lead this process because rural societies are assumed to be insufficiently organised to contribute to their own development.[28]

Through the growing volume of studies on common property resource management it is increasingly recognised that rural societies are not passively waiting to be organised by external agencies. Indeed, viable endogenous institutions for collective action are found all over the world. Research on endogenous institutions managing common resources indicates that the relation between government and the

---

28. Cernea (1985); Bagadion and Korten (1985).

capacity of local communities to organise themselves to pursue their common interests is more complex than earlier assumed. With regard to common or public, natural or man-made resources, governments have a key role in their supply, by determining legal frameworks regulating their ownership and use and by providing the administrative framework through which they are managed. These may have a positive or negative effect on their preservation as well as on a group's collective access and control over a resource. Ostrom (1992b) contends that if viable appropriators' organisations are to emerge, their rights to access, use and potentially, to the exclusion of others need to be sustainable and certain. The state clearly assumes a critical role in providing these conditions. Nevertheless, overly centralised government authority structures are increasingly recognised as having an adverse effect on local initiatives (Ostrom 1992b, McKean 1992).

The attitude of governments towards local communities and their resources is itself central to understanding their inclination towards operating collectively. Popkin (1979; 1993), for example, with reference to local initiatives to construct flood protection embankments in rural Malaysia, explains local collective action as a response to a governmental inefficacy to provide communities with the required public utilities. Taylor and Singleton (1993) place a high value on a community's capacity to organise by itself and contend that by intervening, governments may adversely affect a group's capacity to solve its problems endogenously. A similar position is held by McKean (1992), who notices that while it was earlier assumed that the government was responsible to create institutions and organisations at the local level to manage common resources, it is now often believed that leaving villages essentially alone with the management of their commons may be beneficial from an equity perspective and also in terms of resource sustainability.

Mearns (1995) captures the often problematic relationship between the state and local communities by pointing at the difference between structures of governance and those of government. In the absence of government, either because external rules have not been devised or because weak administrative capacity hinders the enforcement of law and policy, governance — or the lack of it —

becomes entirely a question of endogenously evolved sets of rules. A government's political and administrative strength across regions may vary, from being relatively strong in accessible areas to being almost absent in remote areas. Accordingly there may be strong variations in scope and necessity for local communities to organise and develop their own institutions endogenously. As noticed by Wade (1988), although government institutions may be almost absent at the village level, they may still inspire endogenous institutions:

> It is very doubtful that the 'Union *Panchayat*' at the bottom had even as much significance as its successor, the village *Panchayat* of today. On the other hand, this government structure of committees and special funds, operating at a level not very far above the village, may have provided institutional models for village-based institutions.
>
> — Wade 1988: 27

The following chapters show how government interventions may in some cases trigger collective action and in other cases make it redundant. When the government supply of a public good replaces local provision, one implication is that costs are transferred to the state or to international development agencies. Nevertheless, the fact that the government at one point may take over a responsibility previously carried out endogenously does not mean that a community will forever rely on the state and henceforth become dependent on external assistance (Duyne 1998).

Summary

This chapter showed that participation in collective action should not be excluded but cannot be taken for granted either. According to collective action theory, whether and how a group succeeds in pursuing a common goal, primarily depends on group characteristics, such as size, degree of homogeneity and the level of interdependence among its members. As will be discussed in the following chapters, group size and heterogeneity are of critical importance to understand *how* a group organises to provide itself with a collective good. Yet, collective action theory tends to neglect the context of group action and thus is unable to explain *why*, within the same context, some groups are far better organised and more successful in mobilising resources for collective purposes than others.

This book emphasises the importance of explaining the conditions under which a group is likely to succeed in collective actions, by adding to the analysis of group attributes a systematic review of contextual factors related to three domains: the socio-economic conditions, the agro-ecological situation and the role of the state. Based on these considerations, the following chapter will provide the general contextual information necessary to understand the role of endogenous water management initiatives in rural Bangladesh that will be discussed in Chapters 4, 5 and 6. This will further help appreciate the considerable regional differences in a country which is often assumed to be rather homogeneous. An analysis explaining why collective action is more likely in some areas than in others will be presented in Chapter 7.

3

The National Context

Natural Environment

BANGLADESH covers an area of 144,570 km^2 and occupies one of the most extensive delta systems in the world. The country is located at the confluence of the Ganges, the Brahmaputra and the Meghna, three of the world's largest rivers. It is characterised by vast and fertile floodplains criss-crossed by innumerable rivers and creeks. As illustrated in *fig*. 3.1, with the exception of the hill tracts located in the south-east of the country and two pleistocene terraces, floodplains represent Bangladesh's dominant physiographic feature, making up for 80 per cent of its territory.

The Bangladesh landscape is characterised by its dense river system. Most of the country's 250 permanent rivers consist of tributaries and distributaries of the Ganges, the Brahmaputra and the Meghna. The Ganges, known in the Bangladesh territory by the name of Padma, enters the country in the west, after meandering from the Himalayas through India for about 2000 km. The Brahmaputra, an 1800 km river originating in the Tibetan Himalayas, is known for its last 275 km in Bangladesh by the name of Jamuna. The Meghna is a much shorter river but owes its importance to being fed by the waters of the Meghalaya Hills, an area with precipitation levels that are among the highest in the world.[1] The Ganges, Brahmaputra and Meghna eventually merge into a river known as the lower Meghna, whose average discharge is exceeded only by the Amazon and the Congo rivers (Rashid 1991: 65).

1. The mean annual rainfalls in the Indian Meghalaya plateau are of 11,615 mm (Brammer 1990a).

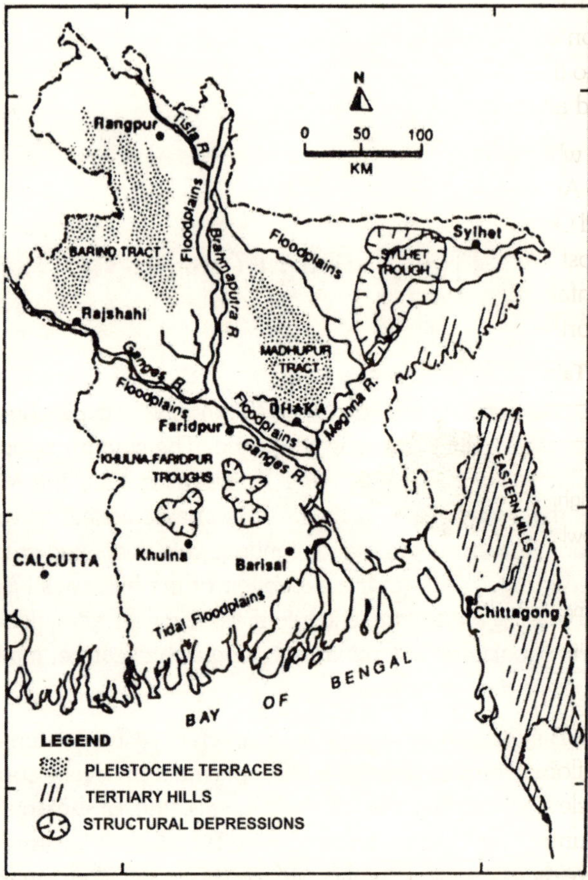

***Fig.* 3.1: Major physiographic features of Bangladesh**[2]

Rivers, canals and streams occupy about 10 per cent of the country's territory and have a total length of over 24,000 km. They constitute an ever present element in the life and culture of Bangladesh. The volume of water carried by this complex river system is immense, for it transports the rains and melted snows of a catchment area of 1,758,000 km^2, of which only 6 to 7 per cent is located within the boundaries of Bangladesh. Their annual discharge of 1,360,000 million m^3 would be enough to form a lake with a depth of over 10.3 m and a size of the whole of the country (Ahmad 1989: 26). The rivers also

2. *Source:* Rasid and Pramanik (1990:816)

carry an immense volume of sediments, which contribute to the formation of temporary or permanent islands and sand bars (called *chars*), to the siltation of riverbeds and canals and to the formation of new land along the coastal belt.

The whole basin's water regime is influenced by a tropical monsoon climate. At least 80 per cent of the annual rainfall occurs between June and October, whereas the months between November and March are almost completely dry. With an annual average of about 1200 mm, rainfall is lowest in the north-west. It is almost four times higher in the north-east, averaging around 4700 mm a year.

Table 3.1: Floodplain Land types by Inundation Level

Land type	*Normal flood depth*	*Proportion (%)*
Highland	0 - 30 cm	17
Medium highland	30 - 90 cm	40
Medium lowland	90 - 180 cm	15
Lowland	180 - 300 cm	9
Very lowland	> 300 cm	2
Water bodies and human settlements		17
Total		100

Source: Brammer (1990a:15)

The floodplain topography is relatively flat but not uniform. The average elevation is of only 4 to 6 m above sea level and 60 per cent of the country has a height of less than 6 m. Minor variations in land height are important because they define the probable level of inundation during the monsoon and thus determine distinct land uses. Based on normally expected annual inundation levels during the monsoon season, farmers as well as agricultural scientists, make a distinction between very lowland, lowland, medium lowland, medium highland and highland. As indicated in Table 3.1, nation-wide medium highland, meaning land annually inundated between 30 to 90 cm, is the most common type of land making up for 40 per cent of the total.

Normal floods are a natural and predictable event, each year inundating about 90 per cent of the floodplain between a minimum of a few centimetres and more than 6 m (Brammer 1990a). Whereas the English language knows only the word 'flood' to designate this complex and diverse phenomenon, the Bengali language makes a basic distinction between benevolent floods, known as *barsha* and damaging floods, known as *banna*. The term *barsha* refers to the normal inundation of the agricultural fields. It is an anxiously expected event by all people, one that recharges the surface water bodies and aquifers and is believed to enrich the fertility of the land through the fine silt carried by river water. It is crucial for *aus* and *aman* paddy, the two most important crops traditionally grown in Bangladesh. A regular *barsha* may inundate all agricultural fields to varying depths but it never inundates the homesteads. Depth, however, is just one criterion to defining the value of *barsha,* which is perceived as more or less beneficial depending on its timing, duration and magnitude. If flooding occurs either earlier or later than the normal time, if it stays for a longer or shorter period than the usual duration and if flood water rises higher or lower than the usual height, it is perceived as abnormal.[3] Based on these considerations, Paul (1984) identifies 27 distinct types of flood situations.

Banna may result from a number of abnormal situations, but independently of its cause, it is always perceived as disastrous, for it damages crops and properties. *Banna* inundates homesteads and may force people to abandon their homes and seek shelter on roofs, trees or embankments. In extreme cases it may cause death to animals and people.

Both for normal as well as for damaging floods, a distinction needs to be made between river floods, flash floods, storm-surge floods, rainfall floods and floods caused by human interference. These flood types affect different regions, though more than one type of flood may occur in the same region. Storm-surge floods are associated with tropical cyclones and hurricanes, which periodically move up

---

3. What is considered normal flood in terms of timing, duration and depth mainly depend on the soil local topography and on the locally specific patterns of human adaptation. In the Jamuna floodplain studied by Paul (1984) normal floods were expected in June-July, to last for not more than four and not less than two months and to have a depth of not more than 8 ft.

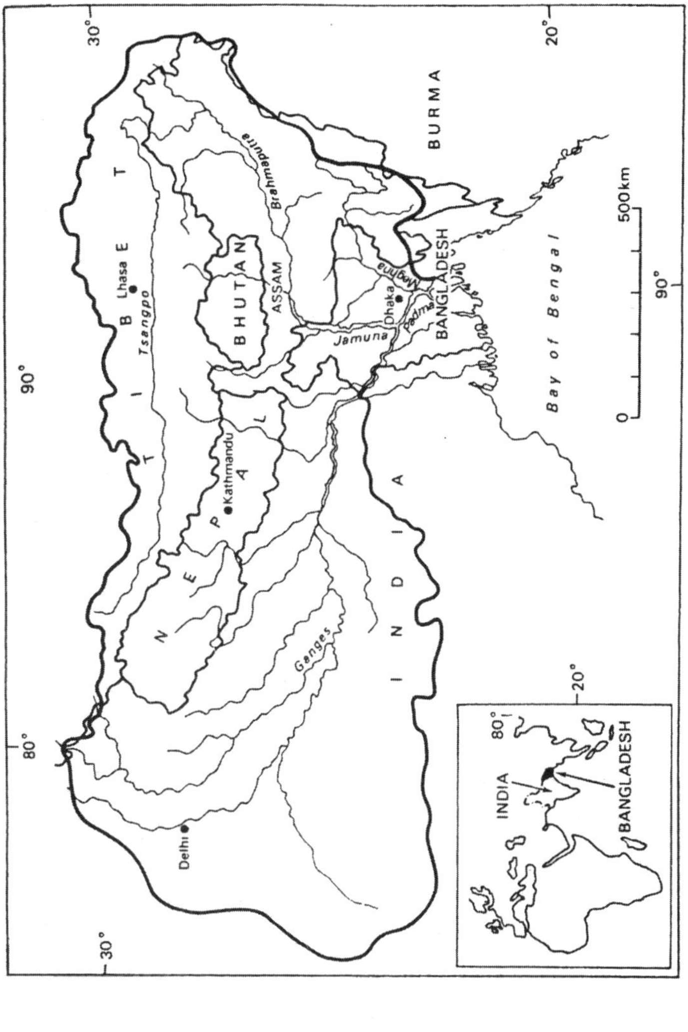

Fig. 3.2: Catchment area of the Ganges, Brahmaputra and Meghna rivers[4]

---

4. *Source*: Brammer (1990a:13)

from the Bay of Bengal and affect the coastal area. They may be very devastating and completely destroy human settlements, infrastructure and crops. In addition they inundate vast areas of crop land with damaging saline water. This type of flood may also kill large numbers of animals and people. The last severe cyclone storm-surge flood occurred in April 1991 and caused the death of an estimated 130,000 people (Hughes *et al.* 1994).

Flash floods are caused by rapid surface water run-off resulting from exceptionally heavy pre-monsoonal and monsoonal rainfalls that may occur in the terraces and hilly tracts inside as well as outside Bangladesh's territory. Flash floods are characterised by a very sudden rise of water levels, usually occurring within a few hours or days. They deposit very large amounts of sediments on the land and within river channels, thus gradually raising river-beds and levee levels.

River floods result from snow-melt in the high Himalayas and by heavy rainfall over the Himalayas, Assam and the Tripura hills. They may or may not be destructive, depending on their timing. They may damage standing crops if they occur too early (June) or too late (after mid-August) and if they occur in combination with flooding caused by rainfall.

Rainwater floods are caused by heavy rainfall within Bangladesh. They are the most common type of flood in Bangladesh. Early monsoon rain may initially accumulate in depressions or drain into the rivers. Once the water levels in rivers and depressions are too high, rainfall water accumulates on the land. As with river floods, whether they are damaging or beneficial depends on their timing and duration.

Floods may be caused or exacerbated by human interference in the environment. For example, river floods may be particularly dangerous if caused by the breach of an embankment. A rainwater flood may become a problem if drainage is hindered by roads or embankments. This may force people to cut them to alleviate drainage congestion in their lands. Floods outside areas protected by embankments may be particularly severe because the water that used to spread over a large territory now concentrates in a smaller area. This is particularly a threat for *chars* (interfluvial lands) and their vulnerable inhabitants. In some coastal areas farmland is deliberately flooded with saline water to facilitate the production of shrimps, which over the last fifteen years has become one of the main export products of Bangladesh.

Whereas people's livelihood in Bangladesh relies on normal floods, abnormal floods go beyond their capacity to cope. The average return period of normal floods is only 2.27 years. Moderate floods may be expected every 4 years, severe floods every seven and catastrophic floods every 33 to 50 years (Hofer 1998). In any one year distinct regions may experience very different flood situations. Some areas may be affected by destructive floods, while others may suffer due to insufficient floods. Moderate to severe floods may cause partial damage to standing crops but are generally followed by a bumper crop, which to some extent may compensate for the loss.

In any case floods are a seasonal and thus in most cases only a temporary concern. During the dry winter months Bangladesh's agricultural production is constrained by water scarcity, which constitutes people's principal concern during the winter and spring seasons. From a climatic point of view, winter would be an ideal season for cultivating a wide range of crops but due to insufficient moisture and irrigation facilities, much land remains fallow. This situation led Boyce (1987; 1990) to argue that the main water control dilemma of Bangladesh is not to check its floods, but to store rainfall and surface water in the months of excess supply for release in the deficit months, a position — as will be discussed later — clearly reflected in people's endogenous water management practices. The multiplicity and seasonality of water control concerns faced by the people in Bangladesh have been overlooked by many scholars and policy makers, who all too often narrowly focus on abnormal floods.

Patterns of Human Adaptation

> *Over the centuries, the cultivators and fisherfolk of the Bengal delta have learned to survive in a setting in which the boundary between land and water is elusive and ever changing. Their accomplishment should not be romanticized: millions of*

> *Bangladeshis today live near the edge of subsistence, while agricultural potential remains unfulfilled. Neither should their accomplishment be under-estimated.* — Boyce 1990: 426

The successful adaptation of human populations to the floodplain environment is clearly reflected by the fact that Bangal has historically been one of the world's most densely populated areas. In 1997 Bangladesh counted a population of 124 million and had 861 inhabitants per km^2. With the exception of a few city-states, it is the most densely populated country in the world. About 80 per cent of its inhabitants live in the floodplains and more than one-third in lands that are less than one hundred years old (Rasid and Paul 1987). The natural environment is characterised by fertile soils and rich water resources. The indigenous production system is primarily based on agriculture and fishery. Thus, this section focuses on Bangladesh's agricultural practices, on the role of fisheries and the characteristics of the homestead.

AGRICULTURE

Even today the livelihood of the vast majority of the rural population depends on farming. Agriculture employs 61.3 per cent of the labour force and contributes to around 46 per cent of the GDP (World Bank 1996). About 67.5 per cent of Bangladesh's soil is suitable for cultivation. The land is fertile and sustains a rich variety of crops. Nearly one hundred crops are currently grown in Bangladesh, of which 32 occupy 96 per cent of the total cultivated area. Rice, Bangladesh's staple food, constitutes by far the most important crop. It covers around 80 per cent of the gross cultivated area and constitutes 75 per cent of the total produce from all field crops (Rashid 1991). Cropping patterns are intricately adjusted to the annual flood regime and primarily depend on the soil's topography.

The growing season extends throughout the year, but may be divided in three distinct though partially overlapping cropping seasons. The first season, known in Bengali as *bhadai* but called by agricultural scientists as *kharif* I starts in late March and lasts until the end of September. It is followed by the *haimantic* season (*kharif* II), which begins in August and lasts until early December. The dry winter season, called *rabi*, covers the period between November and early April. Rice can grow throughout the year under different flooding conditions. Bangladesh's hundreds of rice varieties may be divided in three distinct types known as *aus*, *aman* and *boro*. Besides differing in texture, flavour and yields, they also differ in their growing seasons, habitats and sensitivity to floods.

Aus is the main crop grown during the *bhadai* season. It is commonly sown in February or March at the edge of *beels*, that means in upper middle to lower middle land, which in normal years is not subject to severe flooding. *Aus* paddy greatly depends on monsoon rains, but does not tolerate excessive amounts of standing water. It matures in only 80 to 120 days and is harvested from boats in August and September, when the water in the fields is still 1.5 metres or more in depth. Weeds can reduce its yield by nine-tenth, but even in the absence of this threat yields are quite low, averaging only 330 kg of rice per acre (Rashid 1991: 190).

One of the rice varieties that is doing best in adapting to Bangladesh's floods is known in English as 'floating paddy' or 'deep-water paddy' and is locally called *jali aman*. *Jali aman* is sown in March or April in lands comparatively lower than *aus* which does not require any manure and which starts getting flooded by the middle of May. Depending on the variety, it flowers from August to September and is ready for harvest in November. *Jali aman*'s importance has been decreasing with the introduction of flood control and drainage systems, but it is still cultivated in the vast low-lying areas in the centre of the Bengal delta. *Jali aman* remains the most important crop in areas where monsoon flooding is normally one metre or more in depth, that is, where no other crop can be grown in summer. There are three different categories of *jali aman*, which have distinct capacities to withstand flooding and different growing periods. Shallow flooding varieties grow in 1 to 1.5 metres of water, and medium varieties grow in 1.8 to 2.4 metres of water, whereas deep varieties can grow in water up to 3.6 metres in depth. They have long thin stems, which may reach a length of up to 6 metres. *Jali aman* varieties have the advantage of adapting very well to extreme flood conditions. With rising flood levels they may grow up to 15 cm a day. However, their yields are low. Early and late ripening types only give about 250 kg of rice per acre, and medium types, about 363 kg. Their average yield is estimated to be only 304 kg per acre (Rashid 1991:189).

Both *aus* and *jali aman* are sensitive to drought as well as flood. *Aus* may suffer due to shortage of rainfall at the initial growth stage and be damaged by flood before harvest. *Aman* on the other hand may succumb to excessive floods in its early growth phase or suffer from drought in October in its flowering p eriod. In many areas, to adjust to these risks and ensure that at least one crop is obtained, *aus* and *aman* are sown together. After the *aus* crop is harvested, the *aman* continues to grow and is harvested several months later.

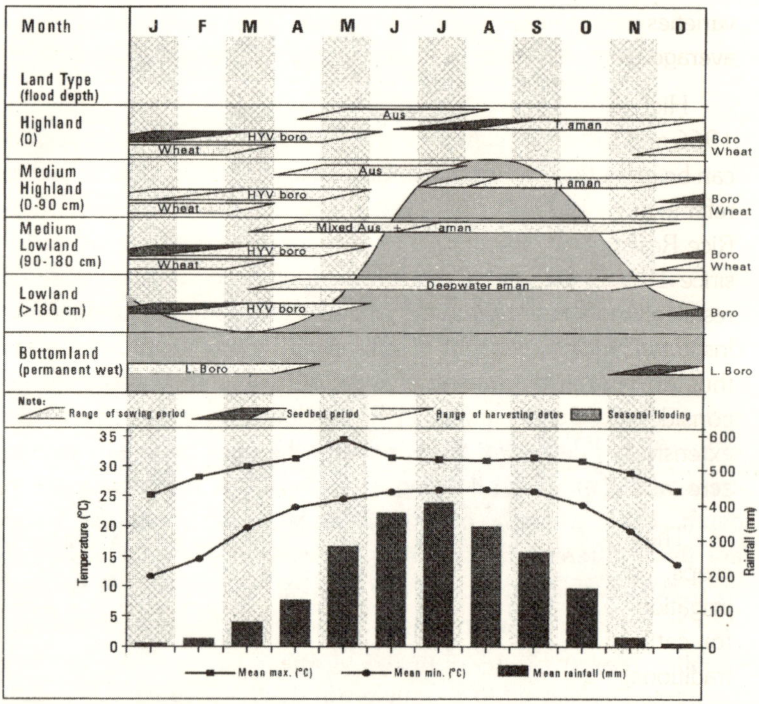

Source: Hofer (1998:7)

***Fig. 3.3*: Rice and wheat crop calendar in relation to seasonal flooding and temperature**

Another important crop grown in the *bhadai* season is jute. Jute was traditionally cultivated in Bengal on a small scale for the production of ropes, paper and cloth. It started to be planted on a large scale in the first half of the nineteenth century, because of the increased demand for cheap packaging material resulting from British industrialisation and world trade. It soon became the country's most important cash crop.

The *haimanta* season is dominated by the cultivation of transplanted *aman* (T. Aman) known as *shail*, the most important paddy crop grown in Bangladesh. *Shail* seedlings are grown in nursery beds between June and July and transplanted five to six weeks later. The nursery beds may then be used to grow *rabi* crops after the second half of September. The *aman* crop is harvested between November and January, depending on the variety. Traditional *shail* varieties may yield up to 816 kg of cleaned rice per acre, though the average yield is about 590 kg of rice per acre (Rashid 1991: 202).

High yielding varieties (HYVs) were introduced in Bangladesh in 1966 and today there are about 26 high yielding *aman* varieties that can be grown during the *haimanta* season. Even these days, farmers refer to them as *bideshi* (foreign) or *IRRI*, the initials of the International Rice Research Institute, where they were first developed. However, since 1970 the HYVs released to the Bangladeshi farmers are the result of breeding and selection at the Bangladesh Rice Research Institute (BRRI). They are shorter than the indigenous varieties and thus do not adjust as well to floods. Furthermore, they are not considered as tasty. Nevertheless, farmers have adopted them extensively for their yields, ranging from 900 to 1360 kg of rice per acre, and are significantly higher than those of any local variety.

The *rabi* season would be an ideal time to grow a rich variety of crops, but due to shortage of water, until the advent of modern irrigation technology, much land remained fallow and was used only for cattle grazing. In most areas of Bangladesh the main crops traditionally grown during the *rabi* season are pulses, oil seeds, sweet potatoes and winter vegetables. Millet, sorghum and barley are still commonly grown in the dry north-west. Of particular importance are red lentils, which together with rice and fish constitute a staple of the

Bengali diet. Most traditional *rabi* crops can grow on residual moisture or require little supplementary irrigation.

Before the introduction of modern irrigation technology rice could not be grown very extensively in the *rabi* season. Winter varieties known as *boro* were traditionally cultivated exclusively in small portions of very low lands in *beel* areas and in the *haor* basin, where surface water for irrigation is plentiful and the depth and length of floods do not allow for cultivation of any other crops. Nevertheless, the dry and sunny winter season is ideal for growing rice. Indigenous varieties of *boro* yield an average of 726 kg of clean rice per acre, which is better than the yield of any other local variety of rice.

Nowadays irrigated high yielding varieties of *boro* are grown in the *rabi* season in most of Bangladesh. Although their taste is not widely appreciated, their yields, which are consistently higher than 1360 kg of rice per acre, make them a very attractive crop, in spite of the fact that, unlike indigenous crops, they are heavily dependent on fertilisers, pesticides and irrigation. Where irrigation has become available, *boro* HYVs have rapidly replaced traditional winter crops, such as various types of pulses and oil seeds. Today the total production of HYV *boro* is almost equal to that of *shail*. This crop has made farmers dependent on agro-chemicals and modern irrigation technology but it also has allowed, particularly over the last decade, for a tremendous growth in rice production.

Bangladesh's traditional farming system reflects a careful adaptation to its flood regimes. With the exception of cropping patterns prevailing in the winter season, agricultural practices have not been modified dramatically by the advent of modern technology. Even today most land is cultivated with very simple implements, most of which are produced locally. The plough, for example, has not changed its form in the last 2000 years. To plough, sow, weed and harvest the land of an average farm of about two acres, a farmer needs two bullocks, a plough, a ladder, a rake, a levelling beam and a variety of sickles and knives. The paddy is blown dry with the help of a winnowing basket made of bamboo. A wooden mallet, known as *dheki*, was traditionally used for husking the paddy. Bangladesh's floodplain land is very fertile. Green manure, ash and cow dung were traditionally used as fertilisers only for the not-regularly-inundated high lands and for cash crops such as sugar cane or tobacco. The remaining soils at most benefited from the manure left behind by the cattle while grazing in the fallow fields, although cow dung is more extensively used as fuel than as manure.

Men and women are both involved in agriculture, but there is a strictly gender based division of labour in family operated farms. Men are responsible for those agricultural tasks that take place in the open fields, which include ploughing, sawing, transplanting, weeding and harvesting. Women are in charge of those tasks that take place in the homestead, which include all pre-cultivation works such as the preservation, selection and germination of paddy seeds and for post-harvesting activities. They assist in threshing and are in charge of parboiling, drying, storing and — prior to the diffusion of mechanised rice mills in the early 1980s — husking the paddy. Women are also in charge of cattle, livestock, poultry and ducks, caring for trees, fruits and vegetables grown in the homestead, making the cow dung sticks and oilcakes (made of mustard seed residuals) used for fuel, fodder or manure. Women are thus heavily involved in agricultural production and are generally well informed about their farm, though in most areas of Bangladesh they are rarely seen in the fields.[5]

FISHERY

Bangladesh's immense inland open water resources host one of the richest and most diverse fish populations in the world, consisting of around 270 species of freshwater fish and 63 species of prawns. According to official estimates, fresh water fish production per unit area in an undisturbed floodplain reaches 693 kg/ha per year, which is the highest in the world (World Bank 1991). The importance of fish in Bengali culture and nutrition can hardly be overemphasised. A fish preparation traditionally accompanies almost every meal and constitutes the main source of animal protein. The annual per capita

5. There are some regional variations in this regard. In some regions of Bangladesh, such as, for example, Banskhali and Galachipa, women can be sporadically seen in the fields, whereas in others, such as Beel Singri and Chaptir Haor, female participation in field tasks is almost inconceivable.

consumption of fish is estimated around 7.3 kg (Ali 1997). The Bengali proverb *mache bhate bangali*, or 'Bengalis are made of fish and rice' is a clear expression of the importance given to fish in Bangladesh.

Fisheries in Bangladesh can be classified into three basic production systems. Inland open water capture fishery, inland closed water fishery (primarily referring to fish culture in ponds) and marine capture fishery. Open water capture fishery, consisting in harvesting naturally occurring and self-reproducing fish populations living in rivers, canals, *beels* and in the seasonally flooded floodplains, is particularly important, making up over 42 per cent of Bangladesh's total fish catch and around 80 per cent of the inland fishery (Middendorp *et al.* 1999).

Inland open water capture fishery involves most households in rural Bangladesh. A basic distinction, however, is to be made between communities of traditional fishers and rural people fishing for their own subsistence or to supplement their income. According to official statistics around ten million people directly or indirectly gain their livelihood from fisheries. However, only a small fraction consists of full-time fishermen. Most traditional fishermen are Hindus, although traditional Muslim fishermen can be found in some areas of Bangladesh.[6] Regardless of their religion, they have a low social status and are generally very poor. Their access to fishing grounds is restricted, because fish resources are officially a state property and have historically been treated as a source of revenue. Fishing rights in rivers, *beels* and canals are generally leased out to rich and influential people who can afford the costly bidding procedures. Traditional fishermen thus generally exercise their profession during the dry season, as labourers or as sub-lettors on very exploitative terms.

During the monsoon season all aquatic areas are connected with the private paddy fields into a single open access resource system from which it is virtually impossible to exclude users. In this season

---

6. For a detailed account of the numerous Hindu castes and Muslim communities involved in professional fishing see Pokrant *et al.* (1997).

surface water systems become a common property resource where all people have the customary right to fish and over 80 per cent of all rural households are involved in subsistence fishing. Free access to this seasonal resource is particularly important for the poor as a source of food and supplementary income in a period of the year offering few, if any other, employment opportunities. It is estimated that around 35 per cent of the total inland open water fish are captured in floodplains, that is, in temporarily flooded paddy fields (Ali 1997:147).

Inland open water capture fishing is practised with over 50 distinct types of nets and other gear and is often linked to a variety of water management practices. In embanked areas at the beginning of the monsoon, people may open up dams or sluices to allow the river water carrying fish fingerlings and spawn to spread over the floodplain. Depending on the context and specific timing, farmers and fishermen may agree on such practices, which may be perceived as functional for agriculture as well as fishing. In many cases, however, there is a severe conflict between these two interest groups, because farmers still have their crops in the field at a time when fishermen would favour overflowing the plains with river water. In general, farmers have the upper hand over the socially and numerically weak fishermen, but in those areas where *beels* are leased by people belonging to the powerful political and economic elite, purposive flooding of paddy fields occurs quite regularly. This has become a pervasive practice in coastal areas, in relation to the highly commercialised and export-oriented shrimp culture.

At the end of the monsoon, water is often purposefully drained to concentrate the fish in a smaller area and facilitate the fishing harvest. In addition nets and traps are installed at drainage points to capture the fish carried by the current. In this case the interests of farmers and fishermen may either conflict or coincide. In some areas, such as the *haor* basin, both groups are interested in drainage, for it allows fishermen to capture the fish and farmers to begin tilling their lands. In other areas, however, farmers would favour retaining surface water for irrigation. They thus try to prevent fishers from draining off surface water. Conflicts of interest between farmers and fishers have been exacerbated with the introduction of less flood-resistant, high yielding varieties of paddy; with the increased water demand during the winter season; and particularly with the introduction of embankments and sluices, which allow for effective control of water levels.

Bangladesh's open water fish resources are in rapid decline. According to official estimates, fish consumption has decreased dramatically, leaving a protein vacuum in the diet of poor people. The reasons behind the depletion of fish resources are several. The abundance of fish has long been taken for granted, leading to indiscriminate and uncontrolled harvest. Government policies and administrative arrangements concerning inland open water fisheries historically have had only one objective, that of raising revenues by leasing access rights to fishing grounds to the highest bidder for relatively short periods of time. These policies have clearly promoted overfishing without any concern about the long-term consequences for fish resources (Farooque 1999; Pokrant *et al.* 1997). Bangladesh still lacks a management strategy or concept for living aquatic resources. In its absence many species already have disappeared or are threatened by imminent extinction. Another factor that has contributed to the current situation is the increased use of agro-chemicals in modern agriculture; agro-chemical residues are considered a major cause of numerous fish diseases (Ali 1997). However, by far the main cause behind the rapid decline of fishes in Bangladesh's inland water systems is related to flood control and drainage works. Indeed, flood control and drainage projects never even considered their impacts on fish, prawns and other living aquatic resources. Embankments are obstructing the seasonal migration of fish, which is indispensable for their feeding and breeding and accordingly systematically destroying their natural habitat. A World Bank study focusing on the impact of flood control and drainage schemes found a nearly total loss of fisheries in some flood control projects, which left scores of fishing villages in decay (World Bank 1990; Minkin 1989). Until recently, such negative impacts of modern hydraulic infrastructure and its socio-economic consequences have not received any attention by policy makers. Whenever the depletion of open water fish resources was recognised as a possible consequence of a water development project, it was considered an externality that could be compensated for by promoting pond fish production. Fish culture in ponds has much increased over the last two decades and still has a tremendous growth potential but it cannot make up for the loss of open water fisheries, because it implies a shift from a common property resource system accessible to the poor, to a private property system controlled by those who are wealthy enough to own ponds.

THE HOMESTEAD

People's remarkable adaptation to the floodplain environment is reflected in the characteristics of the homestead. Homesteads are constructed either on river levees, that is, on land that is naturally 1 to 2 m higher than the surrounding plains or on land artificially raised as much as required to remain above the normal flood level. Each homestead (or *bari*) is generally inhabited by a number of agnatically related families, occupying independent households. The households in the same *bari* generally share a compound used for threshing and drying paddy and for other productive or recreational activities. The homestead also provides a living space for domestic animals, such as goats, cattle, ducks and poultry. The borrow-pits from which the earth is taken are used as ponds which may constitute a joint property of all families sharing a homestead. These are naturally filled with water during the monsoon and are used for bathing, washing, drinking water for animals, irrigating kitchen gardens and increasingly for fish culture. During the wet season, Bangladesh's rural homesteads have the appearance of lush islands. In many areas boats are required to move around. Rafts made of banana stems are seasonally made to transport livestock to grazing grounds. The traditional house is built with mud or straw and has a straw roof. It is surrounded by a wide variety of trees providing fruits (e.g., mango, banana, jackfruit, guava, custard apple), fodder and fuel. These two resources are particularly important during the monsoon, when cattle and livestock must be fed in the homestead compound and fuel becomes scarce. Bamboo trees can be found in almost every homestead and constitute an important source of cash during the lean monsoon season or in case of emergency. Bamboo commerce increases during the rainy season when waterways make its transport cheap and fast. Different types of vegetables are grown throughout the year in most homesteads having sufficient space. These kitchen gardens are an essential source of nutritious food and may provide supplementary income. The homestead is a space controlled by women, who are to a large extent responsible for all activities that take place within its confines. The homestead production system is not only important as an integral part of a household's subsistence; it also provides women with the opportunity to generate independent income and savings.

Aspects of Village Organisation

Bangladesh is a predominantly rural country. Even though the last decade was marked by massive international migration and urbanisation, 82 per cent of Bangladesh's 124 millions inhabitants continue to reside in the country's 68,000 villages (World Bank 1998). Depending on the local topography, a village may consist either of several dispersed *paras* (hamlets) or of a single, densely populated area of land.

For most Bangladeshis the village is an important reference point and unit of identification. However, neither in recreational nor in productive domains do people belonging to the same village act as a corporate group. There are few social and economic activities, which are particularly linked to and confined by village boundaries and there is no social event in which the village population as a whole participates. Interlinkages between villages often overshadow intra-village solidarity, making their boundaries socially 'elusive' (Bertocci 1970). Bangladesh's villages are organised hierarchically according to patrilineal and patrilocal rules. Each village is inhabited by several patrilineages, known in Bengali as *gosti*. Families belonging to the same *gosti* are likely to share the same *bari* (homestead); but they always constitute separate units of production and consumption. The socio-economic status of households belonging to the same *gosti* may vary considerably. Ideally, the more wealthy and powerful households provide to their poorer relatives patronage and some sort of assistance in case of crises in exchange for political and factional support. In reality this is not necessarily the case. The high incidence of poverty, particularly affecting abandoned women and their children, indicates that kinship is not a very effective social security system.[7] Several village studies hint at a gradual attenuation of kinship ties in rural Bangladesh and that primordial loyalties are being superseded by market-based transactional relations.[8] Strong economic differences within a *gosti* may also induce wealthier members to negate common ancestry with the weaker segments of their lineage and lead to the expulsion of poorer households from the *bari* or even the village (Duyne 1987; Jansen 1987).

Several *gostis* together constitute a *samaj*. The word *samaj*, literally means 'society.' The *samaj* is an indigenous social institution through which villagers enact and enforce the normative rules governing their lives. People belonging to the same *samaj* are supposed to be bound by reciprocal obligations towards each other, for example, in times of life-cycle rituals and crises, such as birth, marriage or death. A village may consist of one or more *samaj* (Adnan 1988; Bertocci 1970).

One or more male members of a *samaj* are recognised by its members as their *matabar* or *pradhan* (leader). Advanced age, a honourable moral conduct and wealth are the principal qualities required to be recognised as a *matabar*, a position which nevertheless is often hereditary. Village *matabars* are vested with the task of ensuring that people conform to a morally proper code of conduct. They are members of the *salish*, an informal rural court before which disputes are settled and adjudicated. Individual *matabars* as well as other wealthy villagers may try to strengthen their social status and prestige through public donations of land or cash, for example, for the construction of schools, *madarsas* or mosques. If they have access to the Union Parishad or to the Upazila bureaucracy, they may try to use these contacts to strengthen their power position as brokers between state provided development resources and the village. *Matabars* are generally not associated with efforts to mobilise and

7. The burden of poverty falls disproportionally on women. The most disadvantaged are female headed households, which constitute about 9.7 per cent of all households. According to BBS (1996), 45.3 per cent of the households headed by women are among the hardcore poor experiencing chronic food shortages.

8. Arens and van Buerden (1977), Hartmann and Boyce (1983).

administer local resources to enhance their village's socio-economic development. None of the village studies reviewed by Adnan (1988) attributes to *matabars* any role beyond solving local disputes through the *salish* and ruling over the moral conduct of the villagers. Within this domain they do not appear to act very equitably. Quite to the contrary, many village studies emphasise that *samajas* in current reality do not have much in common with the normative ideal and often use their power position to launch their political and socio-economic aspirations in broader areas.

Agrarian Structure

Although most households involved in farming nowadays supplement their income from agriculture with off-farm employment, the majority of the people in rural Bangladesh still depend directly or indirectly on agriculture. Cultivable land is the country's most valuable resource and land ownership is generally considered the prime indicator of a household's socio-economic status.

Bangladesh is characterised by an unequal distribution of land resources and by a high proportion of landless and land-poor households. According to the 1991 Population Census, about 45 per cent of the rural households do not own any land other than their homesteads. The 1996 Agricultural Census indicates that 27.9 per cent of the farming households hold less than 0.5 acres of land.

The typical Bangladeshi farm is small. As indicated in Table 3.2 and 3.3, farms of less than 2.5 acres account for almost 80 per cent of the total number of farms and over 48 per cent are less than one acre. By Bangladeshi standards, a farm of more than 7.5 acres is

Table 3.2: Type of farms by size and land area controlled

Type of farm	*Owned land (acres)*	*% of farms (1984)*	*% of farms (1996)*	*% of land controlled* (1996)
Small	0.05 - 2.49	70.3	79.3	29
Medium	2.50 - 7.49	24.7	18.2	45
Large	7.50 and above	5.0	2.5	26
Total	-	100.0	100.0	100

Source: 1984 Agricultural Census (GOB/BBS 1987) and 1996 Agricultural Census (GOB/BBS, unpublished)

Table 3.3: Bangladesh's land tenure system at a glance (in acres)

	All-holdings	*Small Holdings (0.05 -2.49)*				*Medium Holdings (2.50-7.49)*		*Large Holdings (7.50 and above)*		
		0.05-0.49	0.50-0.99	1.00-1.49	1.50-2.49	2.50-4.99	5.00-7.49	7.50-14.99	15.00-24.99	25 and above
Size of holding (% of total holdings)	100	27.9	20.3	14.8	16.3	14.0	4.2	2.1	0.3	0.1
Land tenure										
Owner holdings (% of total holdings)	61.5	76.3	56.2	55.9	53.0	54.3	61.8	63.6	76.3	75.0
Number of Tenant holdings (% of total holdings)	1.1	1.8	1.1	0.9	0.8	0.4	0.2	0.4	-	-
Number of Owner cum-tenant holdings (% of total holdings)	37.5	21.9	42.6	43.2	46.3	45.2	37.8	36.0	26.3	37.5
Owned area self-operated	78.4	80.3	71.5	73.8	74.9	77.5	82.4	85.1	88.8	90.0
Rented in area	21.6	19.8	28.5	26.2	25.1	22.6	17.6	14.8	11.2	10.0

Source: Agricultural Census 1996 (GOB/BBS Unpublished)

considered large. This category, however, accounts for only 2.5 per cent of the total number of farms.[9] As may be expected by considering that over the last decade the population of Bangladesh has grown by more than 20 million, farms in Bangladesh are getting ever smaller. The proportion of large farms has halved and the percentage of small farms has increased by 9 per cent.

In aggregate terms, most land in Bangladesh is owner-cultivated. As indicated in Table 3.3, 61.5 per cent of all farmers cultivate only their own land. Farmers cultivating some land in addition to their own constitute 37.5 per cent of the farms, whereas the category of pure tenants is rather insignificant (1.1 per cent). The practice of renting in or out land is relatively common among all landholding categories, though owners of 0.5-1 acre holdings cultivate the highest proportion of leased land in relation to their own land.

Temporary tenancy terms vary depending on the region, the cropping season and whether the owner is an absentee landlord. The contract generally lasts for one cropping season. Neither the owner nor the tenant has any reciprocal obligations to extend the leasing agreement beyond that season. A landowner may decide whether or not to cultivate the land himself based on a number of factors. For example, he may keep his land during the more profitable and less risky *boro* season and lease it out for the *aman* season. The decision may also depend on whether the land is near to his residence, whether he finds a suitable tenant and whether in that particular season he may count on other sources of income or employment. Sharecropping (*borgha*) whereby the tenant passes to the landlord 50 per cent of the harvest is the most common form of tenancy. Traditionally under this type of agreement the cultivator has to bear the full costs of

9. A distinction needs to be made between the *area of land* under tenancy and the proportion of households engaged in lease relationships. The (still unpublished) data thus far available of the 1996 Agricultural Census only gives figures related to the proportion of holdings engaged in leasing land. Nevertheless, Rahman (1989), based on an analysis of earlier official surveys that did provide data on this issue, shows that the tenancy involves a relatively small portion of land but a large number of households. This makes him conclude that the importance of tenancy in Bangladesh lies in the sociological rather than in the production sphere (Rahman 1989:20).

production. Nowadays, in the case of the more capital intensive irrigated agriculture, landlords may be bound to pay 50 per cent of the irrigation costs, for it would be impossible for a tenant to make a positive margin if they had to make up for all the costs of production themselves.

According to the 1960 Agricultural Census,[10] sharecropping accounted for 89.4 per cent of the total rented land. Based on an analysis of numerous micro-level studies, Rahman (1989) concludes that there are no reasons to assume that this situation has changed substantially. This view, however, is not entirely confirmed by a national study conducted by the International Fertiliser Development Centre (IFDC), which showed that though sharecropping continues to be the dominant form of tenancy, the importance of fixed rents in cash or kind has gained considerable importance. IFDC's survey covering 3,000 households in 47 different locations throughout the country, indicates that around 30 per cent of the land is cultivated by tenants. Out of this, 29 per cent is taken in on a cash rent arrangement, 64.4 per cent is sharecropped and 6.6 per cent is mortgaged (Jahan and Sanyal 1993). Fixed leases in cash or kind may appear at a first glance somewhat more attractive, but are generally less appreciated by tenants, for they imply that they have to bear the full risk themselves. In fact, in case of crop damage due to flood, pests, drought or some other natural hazard, tenants are in principle still expected to pay the land rent. Land mortgage, in Bengali known as *bondhak,* relates to the transfer of land from the owner to a moneylender in exchange for a loan. The moneylender is allowed to use the land according to his own wishes and keep it until the full loan is being repaid. Among marginal landowners, the mortgage of land is often a first step towards losing the land completely. The owner may never be able to pay back the loan and therefore be forced a few years later to sell it below the market price.

The agrarian structure of Bangladesh and the failure of national development policies to recognise its nature and its effects on equity

---

10. Cit. in Rahman (1989:19). The 1984 and 1994 Agricultural Census do not provide data on tenurial arrangements.

Table 3.4: Land types and cropping intensity by land-holding categories (in acres)

	All Hol-dings	*Small Holdings (0.05 -2.49)*				*Medium Holdings (2.50-7.49)*		*Large Holdings (7.50 and above)*		
		0.05-0.49	0.50-0.99	1.00-1.49	1.50-2.49	2.50-4.99	5.00-7.49	7.50-14.99	15.00-24.99	25 and above
Size of holding (% of total holdings)	100	27.9	20.3	14.8	16.3	14.0	4.2	2.1	0.3	0.1
Classification of land (% of operated area)										
Highland	34.6	54.2	40.7	37.0	34.2	32.8	34.3	29.9	29.4	19.0
Medium land	46.2	32.9	43.1	45.6	47.3	47.6	46.4	47.9	45.2	50.9
Low land	19.2	12.9	16.2	17.4	18.5	19.6	19.2	22.2	25.4	30.4
Frequency of land use										
Single cropped	30.9	21.1	21.2	23.6	25.8	30.6	35.1	41.0	49.6	58.6
Double cropped	59.2	67.2	67.5	65.6	63.7	59.4	55.4	50.3	42.8	37.1
Triple cropped	9.9	11.8	11.3	10.7	10.5	10.0	9.5	8.7	7.5	4.3

Source:Agricultural Census 1996 (GOB/BBS Unpublished)

and productivity have been pointed out by many scholars as a major impediment to agricultural growth.[11] This position is supported by studies indicating an inverse relationship between farm size and land productivity. The 1996 Agricultural Census data appear to support this theory. As indicated by Table 3.4, marginal holdings of less than 0.5 acres of land have the highest proportion of triple cropped land as opposed to the largest holdings, which have the lowest. Whereas only about 21 per cent of the land held by marginal farms is single cropped, the percentage of single cropped land among large holders amounts to 58 per cent of their holdings. Nevertheless, it needs to be kept in mind that large holdings are more common in the north-east and north-west of Bangladesh, that means in regions where the agro-ecological conditions do not allow multiple cropping. Large holdings' high proportion of lowland supports this interpretation.

As early as the 1970s rural development programmes, with their strong emphasis on the formation of cooperatives through which they channelled state subsidised new technologies and credit, were designed on the assumption that rural Bangladesh was characterised by an undifferentiated peasant society. This has historically allowed the powerful rural elite to capture the benefits of these programmes without necessarily making a productive use of the resources.[12]

The agrarian structure and the relatively unequal distribution of power, combined with a vertical segmentation of the rural society, is considered by many social scientists as a major obstacle for cooperation. This is the position held, for example, by de Vylder, according to whom:

> . . . co-operation is a difficult thing in an essentially competitive, hierarchical society where those who have the most to gain from co-operation lack the necessary means to set the projects in motion and where those who do have the means find it more rewarding to channel the investible surplus in more individualistic, often speculative, directions.
>
> — de Vylder 1982: 136

11. Januzzi and Peach (1980), Boyce (1987), de Vylder (1982).
12. Blair (1974 and 1978), de Vylder (1982).

The effects of Bangladesh's agrarian structure on water control development has also been pointed out by Boyce (1987) as the main cause of Bengal's 'agricultural impasse'. According to Boyce there are several water control systems which would enhance agricultural production and could be initiated at the local level. However,

> Possibilities for labour-intensive water control development projects, which could make use of surplus labour in the slack season, are undermined by landowner disinterest, attributable not only to the formidable difficulties of cooperation among themselves, but also, it is hypothesized, to their reluctance to employ large numbers of agricultural labourers at a single work site for a sustained period for fear of potential repercussions. — Boyce 1987: 247

Even if this view is logical, it needs to be underlined that such attempts to understand rural Bangladesh in terms of Marxist analysis have contributed to a general pessimism about people's capacity to solve their own problems at the local level without the support of external agencies and thus to overlook recurrent instances of collective action. In addition, analyses of Bangladesh's agrarian structure and its implications for agricultural growth have had a tendency to overestimate and misinterpret the importance of tenancy. As indicated in Table 3.3, tenancy is not exclusively associated with a transfer of land from the rich landed rural elite to landless tenants but basically involves all categories of farms. Accordingly, unfavourable tenancy conditions may not be interpreted only as an expression of the rich exploiting the poor.

Another issue that has not gained sufficient attention from social research on rural Bangladesh is the role of the non-agricultural sector and its effects on agriculture in general and on collective action in water resource management in particular. All over the country, though with significant regional differences, there is an increasing economic diversification and a growing number of people involved in off-farm employment. At present, regardless of socio-economic status and ownership of agricultural land, hardly any household relies on a single source of income. Among the poor, diversified sources of income reflect the desperation and creativity of people who cannot afford to be unemployed and who therefore combine petty earnings from a wide range of fluid income sources.

International migration has grown dramatically over the last decade, making remittances from abroad an essential element of both the micro and the macro-level economy. The reliance on income sources not directly linked to agriculture is important for all categories of households indicating the increasingly untenable nature of single income sources. Access to non-agricultural employment may have many repercussions on the agrarian structure. Gardner (1995), based on research in Sylhet District, Bangladesh's region with a historical tradition of international migration, shows that earnings from abroad are often invested in land and may therefore increase polarisation. This may be changing, as international migration increasingly involves also the poorer segments of rural society.

The role of non-agricultural activities may reduce landowners' direct involvement in the management of the farm and thus affect the tenure system. A study carried out in the context of a large-scale irrigation system in the south-east of Bangladesh, found that only 18.5 per cent of the local population reported cultivation as their main source of income. Thanks to the proximity of Chittagong (the country's second largest city), income derived from business, labour, employment in the public or private sector and remittances from migrated family members were generally more important for people's livelihood than agriculture. Landlords were increasingly interested in leasing out their land but due to the low profitability of agriculture (related to the high local labour wages) and the shrinking number of suitable tenants, they were forced to improve the tenancy conditions or to cultivate the land themselves with the help of labourers. The strong reliance on non-agricultural sources of income had a clear impact on farmers' willingness to participate in the management of the irrigation system. In fact, they preferred to pay a commercial price for the water in exchange for a reliable service rather than to directly participate in O&M. Access to non-agricultural employment opportunities for the poor decreases not only their economic dependence on agriculture, but also their social dependence on landlords and may have a tangible impact on wages.

These observations show the complexity and diversity of rural Bangladesh and accordingly the importance of analysing social phenomena, such as local water management practices, in terms of the specificity of the local context.

4

The Role of the State in Water Resource Development

Hydraulic Infrastructure and Water Management Prior to Modern State Intervention

THE concern of Bengal's inhabitants with control over water resources to create a safer and more productive environment dates back to ancient times. Archaeological evidence suggests that canals and ponds have been excavated to provide earth to build dikes for more than a millennium (Haque 1997). However, there is little historical documentation about the role and organisation of water resource management prior to the British period. According to Sir William Willocks, a British irrigation engineer who delivered a series of lectures on this subject in Calcutta in 1923, ancient Bengal had once possessed hydraulic infrastructures comparable in scale and sophistication to those of the Nile and Mesopotamia. Willocks maintains that the Chola kings, who ruled over Bengal about 2000 years ago, inherited their hydraulic skills from Egypt and Babylonia:

> These ancient kings and people came from the north-west and must have had with them many experts from the Euphrates and the Nile; for the irrigation of Babylonia and Egypt was not practised in a corner. The whole world knew about it.
> — Willocks 1930: 5

Willocks passionately describes what he calls Bengal's ancient system of 'overflow irrigation,' characterised by a dense network of artificial canals, fed by water from cuts in the river banks. The aim of these overflow irrigation systems, which allowed for irrigation of more than seven million acres of land, was to 'mingle in the fields of standing rice, the fertilising and health giving waters of the rivers in flood with the poor water of the monsoon rainfall' (Willocks 1930: 5). These water management practices renewed soil fertility, filled storage tanks, re-charged water supplies and also played a central role in checking malaria. Willocks believed that state organised overflow irrigation was disorganised and began to decay after the Mogul conquest.

Willocks did not get much recognition from his contemporaries. The system he described, however, makes sense and to some extent is still reflected in people's current water management concerns and practices. Nevertheless, there is no proof that overflow irrigation was organised by a centralised state authority. This is the conclusion reached by the historian Habib (1963), who found that the substantive revenue literature of the Moguls is characterised by a complete silence on the subject. Thus, if prior to the Mogul conquest, water management was administered by a centralised authority, this was no longer the case under the Mogul rulers.

The absence of a state agency in charge of water resource management does not mean that people and their rulers were passive in this domain. Willocks describes how *zamindars*, after having lost support from the state, themselves maintained the dikes and canals related to the ancient irrigation system, which were erroneously believed by the British authorities to be nothing else than flood protection works. Accordingly,

> . . . zamindars and tenants made secret breaches in the banks and irrigated their lands when they could. These breaches were considered by the authorities as breaches made by uncontrolled floods of rivers and the Government set itself to put an end to such discreditable occurrings. It never seems to have struck anybody that these breaches were made secretly by the peasantry for irrigation.
>
> — Willocks 1930:22

The description of this practice is particularly interesting, as it is still going on at present and like in Willocks' day, continues to be misunderstood by the concerned authorities.

The role of local *zamindars* in the development and maintenance of hydraulic infrastructure is mentioned in numerous District Gazetteers. The author of the District Gazetteer of Rajshahi District, for example, notices the importance of hydraulic infrastructure developed during and after the Pala period:[1]

> . . . there are many large tanks which testify to the munificence either of these kings or of their subjects. One of the remarkable feature is the number and size of old tanks which are scattered over the area. There are two classes of tanks, which perhaps represent two distinct settlements. The older and larger tanks are rectangular in shape, the longest sides running north and south and probably represent an ancient Hindu civilisation. The second series of tanks are square or have their longest sides running east and west and probably represent a more recent Muslim colonisation.
>
> — Siddique 1976: 69

Siddique (1976) does not mention how these tanks were used and managed, but it is well known that tank irrigation played an important role in traditional agriculture in the semi-arid Barind tract, where the scarcity of rainfall makes supplementary irrigation an imperative for the cultivation of *aman* (Rashid 1991).

Zamindars were also active in the development of flood protection works. This is mentioned in the District Gazetteer of Sylhet with a vivid description of the severe floods that occurred between 1785 and 1795:

> Under local rule, the Surma[2] had been protected by a dyke; but this work had been allowed to fall into disrepair and in 1790, Mr. Willes constructed an embankment for a length of about one hundred miles along the side of that river. Six

1. The Pala reigned over extensive parts of Bengal for about 450 years starting in AD 750. Their reign was based in today's underdeveloped semi-arid Barind tract, in the north-west of Bangladesh.
2. The Surma, the Kushiyara and the Manu are three important rivers crossing through Sylhet district. They constitute the main tributaries of the Meghna.

> years before the Supreme Government had sanctioned the expenditure of Rs. 8,000 on the repair of bunds along on the Surma, Kushiyara and Manu, and dykes *were also kept in repair by native landholders at their own expense.* In 1814, the idea seems to have gained ground that these bunds were of comparatively little use. They were accordingly allowed to fall into disrepair and at present day the only one maintained is a dyke along a portion of the Surma river.
>
> — Rizvi 1970: 39-40, emphasis added

The role of *zamindars* in the construction of river embankments also has been noticed near Comilla, in eastern Bangladesh. To finance their construction local farmers were apparently charged a long-term levy consisting of a part of their harvest and they were also forced to maintain them (Hossain 1994).

The British took little interest in water management in the Bengal province and allowed a further deterioration of its indigenous hydraulic infrastructure. The early British rulers were traders and sailors associated to the East India Company who knew nothing about irrigation and never even tried to understand the nature and function of Bengal's ancient canal system (Willocks 1930: 21).

With regard to flood protection works, almost all embankments that were constructed under British initiative were related to the development of roads and railways. These did not keep into account local hydraulic regimes and may be considered the first major interference with natural drainage patterns, to which they caused significant damage. Willocks (1930) accused the British infrastructure of obstructing the natural water flow and depriving the country of the rich red flood water. The belief that British infrastructure contributed to a reduction of soil fertility as well as to the disastrous malaria epidemics in numerous regions of Bengal in the late nineteenth and early twentieth century is also shared by Boyce (1987) and Hughes *et al.* (1992).

Another factor that may have led to the neglect of ancient hydraulic infrastructure is the revenue and land tenure system institutionalised in Bengal under British rule. Whereas the Mogul revenue system still provided some incentive to maintain the ancient hydraulic infrastructure, this was not the case for the system imposed by the British, which led to absentee landlordism, agricultural stagnation and to a gradual abandonment of that infrastructure. The new land tenure system further strengthened a social structure that is considered by several scholars a major impediment to collective action, indispensable for the management of water control systems. According to Mukerjee, for example:

> Throughout Bengal the breach between superior landlords and the actual tillers of the soil ... has led to the neglect of the numerous tanks on which agriculture had formerly depended. The *zamindar* and the higher grade of patni holder have ceased to spend anything for the cleaning of the tanks, which in consequence have been silted up or choked with vegetation, while the cultivators themselves and the tenure-holders of the lowest grades, whose interests it is to see that the tanks are kept in proper condition, are too poor to do anything unaided. — Mukherjee (1933)[3]

The British did not take any interest in promoting irrigation in Bengal, where the level of government irrigation investment was lower than in any other province of British India. Contrary to other states of the Indian subcontinent, where the creation of powerful irrigation departments dates back to the colonial period, the British rulers left Bengal in 1947 without having established any government authority in charge of water resource management (Boyce 1987).

The neglect of irrigation was perhaps partly determined by the belief that the province's plentiful rainfall rendered irrigation superfluous. Accordingly, it is generally assumed that hydraulic infrastructure did not play a significant role during and immediately after the British colonial period and that the agricultural production system was until recently based on individual rather than collective efforts to manage water resources. According to Rasid and Paul, for example:

3. R. Mukerjee, (1933). Land Problems of India. London: Longmans, Green, p. 93, cit. in: Boyce (1987:165).

> . . . in the absence of adequate flood protection structures, the inhabitants of the floodplains in Bangladesh have developed a series of indigenous or traditional adjustments to floods over a period of many generations. Most of these adjustments are undertaken on an individual basis and are therefore considered as private; to contrast them with large-scale public flood control projects. . . . Some of these measures involve adjustments of indigenous structures to floods, whereas others are related to agricultural practices.
>
> — Rasid and Paul 1987:156

The role of irrigation in traditional agriculture cannot be overlooked. Even today, indigenous technologies account for about 12 per cent of the total area under minor irrigation systems (IIMI 1995). The most well-known technologies are three water lifting devices known, depending on the region, as *Seuni* (or *kechi*), *lata* and *dhone* (or *kone*). The *seuni* is a swing basket or scoop operated by two to four persons, who swing it on ropes tied in its four corners. Up to 8.5 cubic metres of water can be raised in an hour with this simple implement. The *lata* raises water from shallow wells. It is a bamboo pole, operating on a fulcrum at the outer end of which another smaller bamboo is tied, which has a can tied to the lower end. The bigger bamboo has a weighted lower end, so that when it is raised and left free the can is dipped in the water and raised. It can lift about 2.5 cubic metres in an hour. The *dhone* is the most widely used indigenous irrigation technology. It has the shape of a three metre long canoe open at one end. The open end is bent towards the surface water source, while the other end is being held up by a bar attached to a pole on a fulcrum with a weighted end towards the land. It is operated by stepping on and off the *dhone,* which dips the closed end in the water. In an hour it can raise up to 63 cubic metres of water, making it by far the most effective indigenous irrigation technology. *Dhones* can be found in most areas of Bangladesh, but are particularly common in the north-west and in the *haor* basin.

In spite of the recognised role of traditional irrigation, little is known about related organisational practices. Given the fact that the above described water lifting devices are owned by individual farmers, it is generally assumed that they do not require any organised action.

Boyce (1987), for example, recognises that water control is to a significant extent a public good and that it cannot be undertaken by cultivators acting alone. However, he believes that:

> Surface water can, to some extent, be utilized by individuals without collective action: traditional lifting devices such as *dhones* and swing baskets, and in some cases simple gravity flow, can irrigate lands adjacent to rivers and marshes. A diesel-powered low-lift pump (LLP), by contrast, requires co-operation among cultivators if it is used to its capacity. [. . .] Tanks, or small reservoirs, require institutional mechanisms for deciding who will contribute land, labour and finance for their construction and maintenance, as well as for governing the allocation and timing of water distribution, a particularly thorny issue in years of water scarcity.
>
> — Boyce 1987: 178

Boyce maintains that water resources cannot be managed adequately by the people themselves due to conflicting interests, factional conflicts and free-rider problems. Boyce sees the large number of derelict tanks and silted canals, which could theoretically be restructured through labour intensive local initiatives without state investments as proving the incapacity of people in rural Bangladesh to cooperate on collective projects.

The views prevailing about the role of water management practices prior to modern state interventions are not based on solid data. It appears, however, that over the centuries there has been a gradual move from water resources being managed by the state to a devolution to local rulers. Due to inadequate colonial policies, these would have failed to sustain public hydraulic infrastructure, gradually leading to atomised individual rather than collective water management practices. Bangladesh's rural social structure is so widely believed to be an impediment for collective action, that evidence of organised water management practices has been systematically ignored. Nevertheless, as will be shown in the next chapter, collective water management practices are relatively common all over Bangladesh and sometimes play a pivotal role in the local production system.

The Development of Modern Flood Control, Drainage and Irrigation Systems

As mentioned earlier, British rulers did not establish any authority in charge of water resources management. Only in relation to a number of severe floods that marked the early 1950s, did water development enter into the agenda of the then government of Pakistan. In 1957 the government requested the United Nations to investigate water management options for dealing with floods. The mission diagnosed a shortcoming of hydrological and engineering data needed to plan structural interventions and recommended establishment of a government corporation with the responsibility for water and power development. This led to the creation of the East Pakistan Water and Power Development Authority (WAPDA), by which East Bengal's water resources came for the first time in modern history under the control of a centralised state agency. The WAPDA soon got the profile of a construction-oriented agency dominated by civil engineers. With technical assistance from the United Nations it immediately began with the construction of several massive river and coastal embankments despite the fact that hydrological data required still were not available (Haque 1997). In 1962 East Bengal was once more hit by severe floods. These were followed by two UN missions, which basically endorsed the implementation of massive hydraulic works and in addition recommended the preparation of a Master Plan. A San Francisco based engineering firm was contracted for this purpose and came up a few years later with a 20-year Master Plan that proposed the construction of thousands of miles of embankments, several large-scale canal irrigation systems, nearly one hundred polders and numerous water control structures. The declared objective of the Master Plan was to increase agricultural production by protecting the *aus* and *aman* crops from flood, enhancing their yields through supplementary irrigation and reclamation of wetland for cultivation through drainage. The pursued strategy kept into account neither the views of the farmers nor those of academic disciplines other than engineering; it was based on the trust that floods were nothing but a damaging impediment for agricultural growth. The need for controlled flooding and drainage vital for traditional floodplain agriculture and the impact of the planned hydraulic works on fisheries and navigation were systematically neglected (Hughes *et al.* 1994; Adnan *et al.* 1992).

Soon after Bangladesh gained its independence from Pakistan in 1971, the WAPDA was dismantled and the Bangladesh Water Development Board (BWDB) became responsible for the water sector. The post-independence period was characterised by a growing awareness that the proportion of costs to benefits in large-scale hydraulic projects was too high. This led the government and some donors to favour small-scale, low-cost and quick return flood control, drainage and irrigation projects. The Master Plan, however, never was officially abandoned; some of the large-scale projects it proposed went ahead, supported in large part by the massive international aid made available to Bangladesh after independence. The water sector was particularly favoured by international development agencies. A review of the water development sector carried out in the late 1980s under the auspices of UNDP showed that:

> More foreign money — more than one thousand million US dollars by one authoritative count — has gone into water projects since 1972 than into any other single sub-sector of agriculture.
>
> — Lindquist 1989: 2

Funds, consisting primarily of soft loans, were at first made available by the World Bank and the Asian Development Bank, which by the end of the1980s accounted for over 76 per cent of the development assistance to the BWDB. Further, around 70 per cent of the World Food Programme's 'Food for Work' wheat was allotted to the BWDB for the execution of rehabilitation and maintenance of its hydraulic infrastructure. This allowed the BWDB to become one of the most powerful semi-autonomous agencies of the country. The BWDB is a huge bureaucracy dominated by civil engineers and spending around 150 million USD per year. In the early 1990s it employed over 18,000 people, of whom 10.5 per cent were graduate and undergraduate engineers and 0.8 per cent other professionals. Among them there was not a single social scientist, a fact reflecting a remarkable indifference to the social impact of water management infrastructure. The remaining employees are strongly unionised support staff, who have made it literally impossible for a number of attempted institutional reform programmes to trim the organisation to a more reasonable and effective size (Barenstein 1994a; Lindquist 1989).

International development agencies' uncritical support of BWDB began to slow down during the early 1980s. It was realised that in spite of the huge development investments, BWDB's hydraulic infrastructure had only marginally contributed to agricultural growth, in comparison, for example, with the less capital intensive minor irrigation technologies that started to be introduced extensively from the early 1970s onwards. BWDB's large-scale canal irrigation systems officially account only for slightly more than 5 per cent of the total area irrigated, in spite of the fact that they have been allocated about half of the investment in irrigation development (Howes 1985). The institutional capacity and performance of the BWDB started to be questioned by several development agencies for it corresponded almost perfectly to the stereotype of irrigation bureaucracy, which in those years fell under critical scrutiny all over Asia (Chowdhury 1989; Botrall 1983). The BWDB was described as:

> . . . an unbalanced organisation, genetically geared to centralised decision making for the construction of infrastructure. Divisions, subdivisions and 'circles' are imposed on the human and political geography of Bangladesh from an inherent — but carefully nurtured — top-down vision of delivering the goods, whether the subjects like it or not.
>
> — Barenstein 1994a: 89

Indeed the BWDB had historically been oblivious to the rest of the administrative structure of the Bangladesh government and hardly ever established formal links with locally elected government authorities or with other government agencies in charge of domains affected by water development projects. The people who were supposed to benefit from water development projects were rarely consulted and their knowledge about the micro-level hydrological conditions ignored. The BWDB was thus increasingly diagnosed as an organisation biased towards construction with no interest in O&M and in the local and national impact of its infrastructure.

The period when BWDB's institutional weaknesses were uncovered coincided with a new international awareness about the importance of people's participation in operation and maintenance (O&M), related to no minor extent to a growing concern that the state could not afford the burden of managing its hydraulic infrastructure. This led the Government of Bangladesh to recognise that inadequate attention to the institutional framework required for operation and maintenance with effective local participation was a major constraint to water resource development (GOB 1990). The *Fourth Five Year Plan* thus proposed to 'maximise local participation in projects at all stages from formulation to implementation and maintenance' (GOB 1990). This policy declaration was followed by a number of internationally funded projects, which aimed to support the BWDB in translating the GOB recommendations into practice.

The first attempts to introduce some element of local participation in O&M started already before the above-mentioned policy declarations. In the context of surface water irrigation systems the formation of 'outlet committees', 'farmers clubs' and 'pump groups' was not a new concept. These organisations, however, either existed only on paper or were hijacked by the local elite and never really took over any O&M responsibility. In the context of FCD systems, it became a common practice to organise for each regulator so-called 'sluice committees' whose members consisted of a mixture of local bureaucrats and BWDB officials, members of the Union Parishad and a number of influential, but not necessarily representative farmers selected by the BWDB staff. These committees were theoretically in charge of the operation of regulators, that is, they were given the power to control the water levels within their command area. Considering that one single regulator may affect the livelihood of thousands of people with potentially opposed water management interests, it is of no surprise that even from a conceptual point of view this institution was inadequate. In practice, sluice committees often existed only on paper and the respective regulators were managed according to the will of a few influential people whose interests either coincided or diverged from those of the majority. If the local elite had conflicting interests, they would all try to have their way by persuading the *kalashi* to open or close the regulators, often leading to results which would not serve anybody and which had nothing to do with the objectives that justified their construction.

A second phase of official concern with people's participation started in the early 1990s. It was the result of pressure from international development agencies, intellectuals and local NGOs, which in those years began to emphasise people's right to participate to decision-making processes with regard to development interventions affecting their livelihood. A multidisciplinary working group of high-level officials from the Ministry of Water Resources, the BWDB and other concerned agencies was established to prepare guidelines for people's participation, which were officially approved by the Ministry of Water resources in 1995 (GOB 1995). To satisfy the demand for democratic participation as well as O&M requirements, the guidelines stress the need for participation at all stages of the project cycle. In stages related to project identification and conceptualisation, participation refers to the need to consult and actively involve the local population in planning processes. In already existing schemes, participation relates to the need to transfer the burden of O&M to the system users. To this aim the guidelines propose the formation of hierarchically organised water users' organisations. The merit and limitations of this policy document and the process that led to their formulation have gained much attention.[4] At the local level, however, they have had little impact. Attempts to follow the guidelines were made only in the context of a few pilot projects by temporary project staff; they were never applied as a general policy. The pilot projects soon proved that the institutional framework for user organisations was not applicable to the context of Bangladesh and that the BWDB did not have the interest, professional capacity and organisational structure necessary to adopt them.[5] The real situation at the local

4. See Adnan *et al*. (1992); Hanchett (1997), Wood (1995).

5. Duyne 1994a; (1995a; 1997a, 1999a).

level also showed that the emphasis on people's participation in O&M of hydraulic infrastructure may detract from other causes that lead to poor performance. This problem was pointed out by Gisselquist (1991). With his 'taxonomy of failed projects' he shows that only one category of water management systems does not perform as planned due to poor O&M. Indeed, many projects fail to deliver because of a weak concept, others due to poor design and yet another category due to poor construction. Only if a water management system has a valid concept and design and is adequately constructed can it be expected that the system users will engage in O&M.

The critiques against Bangladesh's water development sector that prevailed in the 1980s and early 1990s focused primarily on BWDB's organisational weaknesses, the poor O&M status of its infrastructure and on the economic viability of surface water management infrastructure. These critiques came from experts attached to development agencies and generally included recommendations for new projects and technical assistance or for major policy changes. An entirely different type of critique of the water development approaches pursued by Bangladesh under the guidance of international development agencies gained momentum in the early 1990s in relation to the major floods of 1987 and 1988 that were followed by a highly controversial group of studies collectively referred to as the Flood Action Plan (FAP). In spite of the fact that there was evidence that much of the damage of those floods was caused by the failure of poorly maintained embankments, international agencies once more proposed to find a permanent solution to Bangladesh's flood problem by means of massive embankments among the main rivers. Due to the complexity of such an intervention it was considered necessary to dedicate a number of years to detailed studies about the technical, economic, social, institutional and environmental implications of a project whose initial cost was estimated around ten billion USD. This phase was expected to lead to a number of implementation projects to be executed between 1995 and 2010. The FAP triggered an unprecedented opposition from national and international NGOs, intellectuals, journalists, environmental activists and social scientists. The internationally organised opposition to the FAP led to several independent studies about the socio-economic and environmental impact of flood control and drainage systems and also forced the FAP to integrate in its agenda a relatively thorough review of these issues. Thanks to the significant advance in public awareness about the intrinsic importance of safeguarding the floodplain environment and not last due to the non-availability of the funds required for the massive schemes originally considered, the FAP never entered its second stage. What was left behind of the FAP period is a significant volume of studies that demonstrate the complexity of water resource management in Bangladesh and the adverse socio-economic and environmental effects some hydraulic infrastructure may have.

But how do rural people themselves perceive and respond to state provided hydraulic infrastructure? Several studies that were published in the 1990s in relation to the FAP controversy claimed that rural people are victims of BWDB's hydraulic infrastructure. Their claims are based on a number of well-known cases of failed projects and on interpretations of specific local practices, such as cutting embankments.[6] Adnan *et al*. (1992), based on evidence from one FCD scheme in the north-west of Bangladesh, for example, contends that

> the most visible expression of people's frustration with the counterproductive flood protection structures had taken the form of cutting open embankments to relieve drainage congestion. — Adnan *et al*. 1992: s-2

These conclusions disregard the fact that cutting embankments and river banks — as already vividly described by Willocks (1930) — are indigenous water management practices. The cases that will be presented in the next chapter will further confirm that embankment cuts are not necessarily the impulsive response of frustrated people, but carefully planned and sometimes cost effective water management practices.

Empirical investigations directly addressing the question of whether people want or consider themselves benefited from BWDB hydraulic

---

6. Adnan (1990), Adnan *et al*. (1994), Hughes *et al*. (1994), Haggart (1994) and Rahman (1995).

Table 4.1: Irrigated agricultural holdings by landholding category (in acres)

Item/ Region	All Holdings (0.05-2.49)	Small Holdings (2.50-7.49)			Medium Holdings (7.50 and above)			Large Holdings		
Bangladesh		0.05-0.49	0.50-0.99	1.00-1.49	1.50-2.49	2.50-4.99	5.00-7.49	7.50-14.99	15.00-24.99	25 and above
% of holddings rep. Irrigation	59.5	40.2	62.9	67.6	69.2	69	67.8	67.8	71.1	62.5
% net irrigated area to net cultivated area	43.4	45.4	54.1	51.7	46.8	42.2	39.7	38.2	30.7	17.6
Area under surface water irrigation (in 000)	1,755	72	181	216	354	472	216	179	53	12
Area under ground-water irrigation (in 000)	5,826	203	560	696	1,155	1,640	808	604	129	31
Area under LLPs (in 000)	1,322	48	140	167	267	348	163	139	41	9
Area under STW (in 000)	4,820	160	446	565	941	1,367	689	517	110	25
Area under DTW (in 000)	917	34	99	120	195	252	111	82	18	6
Area under hand-driven devices (in 000)	523	34	56	60	106	145	61	44	14	3

Source: Bangladesh Agricultural Census 1996

infrastructure give a more differentiated picture. An evaluation of the role of water users' organisations in participatory water management, based on interviews with a random sample of 438 members of water users organisations in six BWDB FCD/I schemes revealed that between 86 per cent and 98 per cent of the members consider the FCD/I scheme in their area as generally useful and their lives and properties to be positively affected by it. While the usefulness of the schemes was almost undisputed, on average only 53 per cent of members were satisfied with the services received from the BWDB. There was a general tendency to blame this agency for not operating and maintaining the systems properly, for not responding to the needs and priorities as defined by the people and for not living up to the expectations raised among the local population (Duyne 1997a).

The same questions were addressed by an independent evaluation of an internationally funded water development project in a survey conducted in about 100 villages across the country affected by some seven distinct FCD/I schemes. The study showed that, even by considering the opinion of all people living in FCDI areas and not just the most directly concerned farmers, only 1 per cent of the respondents considered the impact of the BWDB hydraulic infrastructure as negative as opposed to 53 per cent who considered it to provide 'significant' or 'very significant' benefits. The same study, however, also showed that people were aware of some of the negative impacts of FCD/I schemes in particular the loss of fish (Soussan *et al.* 1998).

The fact that rural people appreciate state provided hydraulic infrastructure is important to keep in mind, for it shows that common people and policy makers are equally biased towards rice, similarly insensitive about environmental issues and willing to accept some negative consequences in order to improve conditions for rice production.

Minor Irrigation Development

Ever since independence, Bangladesh has given great importance to the expansion of irrigated agriculture, which was instrumental for the country to become self-sufficient in foodgrain production. The

Bangladesh Water Development Board (BWDB) and the Bangladesh Agricultural Development Council (BADC) were from the beginning the two main government agencies in charge of irrigation development (Chowdhury 1989).

BWDB's endeavours to promote irrigated agriculture primarily involved the construction of medium and large scale canal irrigation systems, a capital intensive technology with only limited development potential in Bangladesh.[7] The development of irrigation was thus bound to rely on minor irrigation technologies, in which WAPDA had shown little interest (IIMI 1995; Boyce 1987).

The start and early growth of minor irrigation systems primarily relied on BADC, a government agency established in 1961 to promote the use of modern agricultural inputs. According to its mandate BADC was responsible to procure, transport, store and distribute seed, fertilisers, pesticides, agricultural machinery and implements as well as to supply, operate and maintain mechanical irrigation equipment. The promotion of minor irrigation technologies aimed at enhancing the production of HYV crops during the dry winter season, that is, disseminating the Green Revolution package that spread throughout Asia from the late 1960s onwards (de Vylder 1982).

The first technology launched by BADC were low lift pumps (LLPs). LLPs draw upon surface water, but can lift the water from greater distances and in greater volumes than are possible with traditional irrigation devices. Initially they were owned and operated by BADC and farmers were required to pay water fees on a unit basis. In the late 1960s, BADC started renting the LLPs on an yearly basis and the farmers themselves had to pay for the fuel. LLPs

7. Bangladesh's topography is too flat for the development of gravity flow irrigation systems. The storage of surface water for subsequent use would require the construction of reservoirs for which there are few suitable locations and which subtract land from agriculture. The expansion of river water deviation is constrained by the fact that at least half the dry season flow must be maintained to prevent the disruption of navigation and limit the intrusion of saline water from the Bay of Bengal. Due to these constraints BWDB's large scale gravity flow irrigation systems were not designed to provide irrigation during the dry winter season, but only supplementary irrigation to *aus* and *aman*. Accordingly, they did not contribute significantly to increasing cropping intensity (Boyce 1987; IIMI 1995; Lindquist 1989).

expanded relatively rapidly during the 1960s and by the early 1970s had overtaken traditional technologies in terms of area irrigated. Their role in irrigation continued to grow during the 1970s. However, the development potential of this irrigation technology was clearly limited by the availability of suitable locations. The focus on any further irrigation development strategy was therefore bound to move away from surface water towards the use of groundwater resources, which by that time were believed to be sufficient to irrigate most of Bangladesh's cultivable land without risks of depletion (Pitman 1993; Saleh and Nishat 1989).

Deep Tubewells (DTWs) were the first groundwater irrigation systems to be promoted by BADC. They are a relatively costly and complex technology that allows for extraction of water from aquifers at an average depth of 75 m. They have a command area of 50-60 acres of land, potentially owned by a large number of farmers. The dissemination of DTWs was thus linked to the organisation of cooperatives, whose members had to make a plot of land available for the well and the pump house and lay out their own water distribution system. The promotion of DTWs was heavily subsidised with farmers being required to pay only nominal annual rental fees. These number grew relatively rapidly from the mid-1970s onwards to 1000-1500 per year.

At about the same time BADC also started a programme of selling Shallow Tubewells (STWs). The STW is a much simpler and cheaper groundwater irrigation technology than the DTW. A STW has a command area of about 20 acres of land at a cost around one-tenth that of a DTW. From the very outset the dissemination of STWs was supported only by minor subsidies, though their purchase was encouraged by agricultural credit banks. They attracted relatively large farmers, who used them to irrigate their own land and sold water to other farmers owning land within their STWs' command areas.

In spite of the heavy governmental support for the expansion of irrigation, the take off was slow and agricultural production failed to grow as expected. By 1980 BADC had only succeeded in fielding around 10,000 DTWs, 35,000 LLPs and 22,000 STWs. Little more than 15 per cent of the total cultivated area was under irrigation, although estimates indicated that over 75 per cent of the land could be irrigated with the help of underground water resources. This poor achievement was a matter of concern and led to a lively policy debate.

The failure to bring more land under irrigation despite heavy governmental support was attributed to several causes. Some studies indicated that less than 30 per cent of the installed capacity of DTWs and only around 50 per cent of that of the STWs were actually utilised. The under-utilisation of DTWs and STWs was attributed to many factors, ranging from technical and institutional deficiencies to social constraints.[8] Many social scientists emphasised that the agrarian structure was a major impediment for the adoption of new technologies or that it would lead to further polarisation.[9] Only if accompanied by an agrarian reform would agricultural development policies lead to equitable growth, a step in which each successive government has shown rather limited interest.[10] Dominant development agencies, however, opined that bureaucratic inefficiency was the main problem and accordingly recommended the reduction of government intervention in irrigation management and the gradual withdrawal of the state from the supply of irrigation equipment and complementary inputs. Given the key role of donor agencies in financing implementation of Bangladesh's development policies, the government had little choice but to follow this advice and to embark into a drastic privatisation programme (IIMI 1995).

The privatisation programme of the minor irrigation sector started in 1980 and was intended as a gradual process. It included a number of aspects, such as the sale of publicly owned LLP and DTW systems to cooperatives or private farmers, import liberalisation, the abolition of regulations related to the type of STWs allowed for import as well as the suspension of rules related to their spacing. The programme was much criticised by progressive circles concerned about the equity implications of promoting privatisation in a highly inequitable agrarian

8. See, for example, Bottrall (1983) and Mandal (1989).
9. de Vylder (1982), Januzzi and Peach (1980), Boyce (1987).
10. Rahman (1994).

structure. There were speculations that private sector control would lead to a monopolisation of ownership of irrigation technologies by large farmers, who would thus add to their advantages of controlling land the additional advantage of controlling water.[11] Small and marginal farmers would thus become increasingly dependent on large farmers. This situation would not only give large farmers the benefits of increased production, but also would enable them to use their control over water to drive marginal farmers off their lands.

Against all odds, from the beginning of the 1980s onwards, agricultural production began to grow at an unprecedented rate. This positive trend was associated with a rapid dissemination of STWs, which currently account for about 52 per cent of the irrigated land (see Table 4.2). This allowed a massive increase in HYV varieties of *boro* rice, which in terms of total production, almost reached the importance of *aman*. The increased role of irrigated agriculture led to major changes in cropping patterns. In many regions *boro* almost completely replaced *aus*, deep-water *aman* and traditional winter crops. Much of the land that used to remain fallow during the dry season came under cultivation. Thanks to the rapid increase of irrigated winter rice, for the first time in modern history foodgrain production grew faster than the population.[12] In aggregate terms, agricultural growth led to increased employment opportunities for rural landless labourers and to the availability of more and lower

Table 4.2: Minor irrigation equipment in relation to area irrigated (1992-1993)

Irrigation technology	*Operating units*	*Area irrigated (ha)*	*% of total area irrigated*
STW	348,875	1,392,198	52.1
LLP	52,217	496,008	18.6
DTW	25,714	436,857	16.4
MOP	134,718	22,413	0.8
Traditional	713,660	323,034	12.1
Total	**1,275,184**	**2,670,510**	**100.0**

Source: IIMI 1995

11. Howes (1985), de Vylder (1982).
12. Rogaly *et al*. (1999), Palmer-Jones (1999).

priced rice, a trend that was reflected by a moderate but tangible reduction of rural poverty (Rahman 1998).

A series of studies by IIMI indicate that STW and DTW ownership has not been monopolised by large farmers. All categories of farms have access to ownership with the largest single group of owners not being large farmers but medium farmers. Over the years small farmers' ownership of STWs has increased markedly. In 1982 small farmers constituted only 14 per cent of the total owners, while in 1994 they were more than 27 per cent. During the same period the percentage of large farmers owning STWs has fallen and that of medium farmers has remained the same. The study found that, depending on the region, farmers owning less than three acres of land owned between 60 per cent and 80 per cent of the STWs. On balance, rather than promoting monopolisation of irrigation equipment by large farmers, provision of inexpensive and appropriately sized equipment made possible by the privatisation policies has expanded the access of small farmers to irrigation systems (IIMI 1995).

Although these positive developments should be appreciated, it is nevertheless important to mention that they are not completely unproblematic. The dissemination of minor irrigation technologies is associated with an increased use of polluting agro-chemicals, which are particularly harmful for the country's fish resources. Fish resources are further threatened by the increased pressure on water and land, which has accelerated the drying-up of many of the country's wetlands. The growth in production of carbohydrates thus has led to a decline in availability of proteins, which are more difficult to replace through imports. The apparently increased profitability of agriculture and the possibility, offered by modern irrigation technologies, of making a profit from land and water resources that used to be available for communal use has often led the powerful rural elite to make legal or illegal claims on common resources that were particularly important to the survival of the most vulnerable people.

Not all experts are convinced that unlimited groundwater extraction is sustainable. In the north-west of Bangladesh the groundwater table, as a consequence of the expansion of STWs and drying-up of wetlands (which have an important role in recharging the aquifers) appears to have fallen dramatically, creating severe drinking water problems. While groundwater irrigation advocates such as Pitman (1993) consider this situation as temporary and regional, for others it is a clear evidence that the agricultural development strategy pursued by Bangladesh is highly problematic (Wood 1996).

Self-sufficiency in foodgrains was a popular political goal, which tended to disguise that food independence has led to technological dependence. All these issues as well as the recent slow down of agricultural growth have led several scholars to express their doubts about the socio-economic and environmental sustainability of the recent achievements.[13] It is beyond the scope of this book to elaborate on each of these critical issues, which nevertheless set the scenario and are basic to understanding people's water management practices.

13. Whether the recent slow down of agricultural growth should be a matter of concern is a controversial issue as indicated by the conflicting interpretations of Palmer-Jones (1999) and Adnan (1999).

5

Canals, Dikes, Cuts and Closures...

The Logic of Local Water Management Initiatives

THIS chapter provides a general description of local water management practices by focusing primarily on their function, benefit distribution, and on the labour and material resources involved. It shows that all over Bangladesh in spite of the unequal distribution of power and productive resources and the relative effectiveness of individual coping strategies, people sometimes are able to overcome collective action problems and provide themselves with valuable common resources.

Collective water management practices in the study areas include the provision, maintenance and operation of twelve distinct types of hydraulic systems, as indicated in Table 5.1. Some of these systems represent indigenous technologies that were developed locally in relation to the water requirements of traditionally planted crops. Other initiatives represent local responses to external interventions and constitute collective attempts to enhance the advantages or reduce the negative impact of state-provided hydraulic infrastructure. Local water management practices may be triggered by the dissemination of modern agricultural technologies or new market conditions. A few initiatives have been taken by groups of people who managed to mobilise some external support (for example, from the Union Parishad or the BWDB) to supplement their own resources. Whereas the majority of the initiatives discussed are entirely endogenous, two have involved the staff of a water resource development project. People recalled these initiatives three years later with a sense of ownership equal to that expressed for endogenous initiatives. They illustrate that external agencies' endeavours to mobilise local people and resources can follow a similar logic, and are not necessarily bound to fail.

The material and human resources involved in local water management practices vary significantly. The construction of a small crossdam or the cut of an embankment may involve only a few voluntary labourers. Works related to the provision of large crossdams or the maintenance of *janghals* (contour bundhs) may, however, require human and material resources that exceed by far the official development budget of the Union Parishad. The number of people and the value of the material resources involved are not necessarily proportional to their impact; small initiatives may affect — positively or negatively — thousands of people, whereas some large ones may affect only a few individuals (Table 5.1).

Irrigation Canals

The most common endogenous water management initiatives are associated with the construction and maintenance of irrigation canals. The construction of irrigation canals at the tertiary level by the farmers themselves is a common phenomenon in Bangladesh. This study only focuses on primary irrigation canals, which are fairly large and the result of well-organised collective actions. Collectively constructed irrigation canals are particularly common in the *haor* area, but they were also found in Beel Singri and Galachipa. These three areas have in common a rather deep water table, which makes the use of groundwater very costly. In Galachipa and Chaptir Haor groundwater is not used for agriculture and the high cost of installing tubewells for drinking water makes people dependent on surface water for domestic purposes as well. The irrigation canals found in the three research sites vary considerably in terms of their history, technical design and associated organisational practices. In Chaptir Haor they constitute a technology essential for indigenous agriculture and most canal systems have been constructed in ancient times. Beel Singri's and Galachipa's irrigation canals are of more recent origin and have been constructed in order to store surface water that has become available because of newly provided sluices and irrigation inlets constructed with the support of a water development project. The financial and labour figures relating to these irrigation canals in Table 5.1 refer to the annual maintenance works, which involve each year thousands of labour days.

ANCIENT IRRIGATION SYSTEMS IN CHAPTIR HAOR

Chaptir Haor is located in the central part of the *haor* basin, an area that remains inundated to a depth of over 4 m for about six months a year. Thus, the only season available for agriculture is the dry winter, when water is scarce. Even today Chaptir Haor's farmers cultivate primarily local varieties of *boro* paddy that are well adapted to this special environment. They are fairly resistant to deep flooding and have a growing period that is about 20 days shorter than HYVs'. The crop is harvested between April and May, just before the whole area gets submerged in deep water. Under good conditions, which depend not only on nature but also on collective human efforts, it is a very bountiful crop. It is sometimes erroneously assumed that traditional *boro* varieties do not require irrigation, an assumption based on the fact that modern technologies can hardly be found in the *haor* basin. However, even traditional *boro* varieties require some irrigation, in particular during the early stage and during the flowering period. Surface water purposefully stored in Chaptir Haor's seventeen *beels*, eleven *kharas* (natural canals) and about 150 *dhobas* (small *beels*), and one *mara nadi* (dead river) is extensively used for irrigation.

The origin of the irrigation systems built around these surface water bodies is very ancient, but they were not built all at once. They have been gradually enlarged to satisfy the irrigation demands of a growing area of cultivated land. While all people in the *haor* are aware of their existence, and most farmers are using them, nobody knows when they were first constructed. In the local dialect irrigation canals are called *palla*. *Pallas* are excavated canals with a depth of 1-1.5 m, a width of 1-3 m and a length varying between 250 m and 1 km. About 400 *pallas* of various sizes and lengths, irrigating an area of over 5,000 acres of agricultural land, were found in Chaptir Haor. In some cases these canals also serve to link *beels* with *dhobas* and *kharas*, which are used as auxiliary reservoirs. Along these *pallas* individual farmers dig little pits from which the water is lifted to the field channels through the indigenous irrigation device named *kone*.

Table 5.1: Overview of local water management infrastructure and practices in study areas

No.	*Type of Initiative*	*Total identified*	*Galachipa (P-55/1)*	*Banskhali (P-64/1b)*	*Chaptir Haor*	*Beel Singri*	*Benefited households*	*Benefited areas*	*Labour days**	*Financial costs* (Tk.)*
1.	Irrigation canals	418	6	-	400	12	20-60	10-40	20-300	minor-30,000
2.	Crossdams	105	35	28	35	7	200-1,000	50-4,000	40-3,000	minor-8,00,000
3.	*Janghals*	92	-	-	92	-	60-6,500	4,000	7,400	minor-45,000
4.	Embankment cuts	20	2	5	13	-	100-7,000	20-14,000	2-12	minor
5.	Closing embankment cuts and breaches	15	1	4	10	-	50-7,000	20-14,000	7-20	minor-4,500
6.	Raising roads and embankments for flood protection	17	1	1	12	3	53-7,000	75-14,000	120-6,000	minor-4,00,000
7.	Drainage outlets/ irrigation inlets	48	25	23	-	-	50-100	10-20	5-7	1,500-4,600
8.	Compartmental dike	1	-	1	-	-	200	140	NA	4,00,000
9.	Salt inlets	3	-	3	-	-	200	140	NA	1,400
10.	O&M of BWDB regulators	11	-	10	-	1	500-2,000	100-350	NA	minor-3,000

No.	Type of Initiative	Total identified	Galachipa (P-55/1)	Banskhali (P-64/1b)	Chaptir Haor	Beel Singri	Benefited households	Benefited areas	Labour days*	Financial costs* (Tk.)
11.	Purchase and management of ex-BWDB LLP schemes	10	-	-	-	10	65-300	60-250	NA	30,000-1,75,000
12.	Clearing *khals* and embankments	2	2	-	-	-	800-13,000	1,700-18,600	400-1000	minor
	TOTAL	742 (100%)	72 (9.7%)	74 (10.0%)	562 (75.7%)	33 (4.5%)	20-13,000	10-18,600	2-6,000	minor-8,00,000

* In case of Chaptir Haor's ancient irrigation canals and *janghals* these figures refer to the financial cost and labour days involved in annual rehabilitation and maintenance works.

While construction and maintenance of field channels are done by farmers individually, *pallas* are maintained collectively. Every year hundreds of labour days are spent by the farmers of Chaptir Haor to re-excavate their *pallas*. All farmers owning land inside the command area of a specific *palla* participate in these yearly maintenance works. There is an informal rule that each farming household should send one person to participate in the re-excavation works for whatever number of days are required. The date and volume of work to be done is decided in the context of meetings called by elderly and well respected farmers. In case of large *pallas* involving farmers of more than one village, these meetings are called by the village *matabars*. The village *chawkidars*[1] are in charge of informing the villagers about the date and venue of each meeting. The re-excavation of *pallas* is not a one-time activity. In fact, at the beginning of the irrigation season, very little excavation is required to make the water flow. Gradually, as the water in the reservoir recedes, *pallas* need to be further excavated. Some farmers are known for having the expertise to decide on the exact gravity to give to an irrigation canal at different stages of the irrigation season. One example of a canal irrigation system in Chaptir Haor is presented below, as Case 1.

Case 1 — Sri Narayanpur Irrigation System

Sri Narayanpur is one of the 32 villages in Chaptir Haor. It is located in the east of the *haor*, near a *beel* named Katua. Besides Beel Katụa, six *dhobas* are located within the jurisdiction of Sri Narayanpur. These surface water bodies constitute important fishing grounds and are extensively used for irrigation. Whereas fishing rights during the winter season are restricted to leaseholders,[2] the water may be considered the area's most

1. *Chawkidars* are village guards employed by the Union Parishad for a minor remuneration. This part-time occupation puts them formally in charge of village security. In practice their role is rather marginal, for they are generally persons with a relatively low social status who within their community have neither the power nor the legitimacy to exercise much control over village security. In Chaptir Haor the village *matabars* have entrusted them with alternative roles, all related to water management.

2. All major *beels* are owned by the government and leased out through auctions to private individuals. *Dhobas* are generally *ejmali* property. As will be discussed in Chapter 6, many *dhobas* in Chaptir Haor are leased out by the *matabars* to generate *panchayat* funds (village funds), which →

important common property resource. For a more efficient use of their surface water bodies, the ancestors of the present inhabitants of Sri Narayanpur built a complex network of irrigation canals consisting of main canals, secondary canals, branch canals and field canals.

All people know the main canals by their names. A fairly large canal, known as *boro palla* (big canal), links Beel Katua with a *dhoba* named Narayan, which has a size of about six acres. From there the water is again distributed through three secondary canals. The total length of the canal system is about 4.5 km. Along the canals there are 66 pits from which the water is lifted to the field channels by means of *kones*. The Sri Narayanpur irrigation system has a command area of 232 acres belonging to a community of about 160 farmers. During the monsoon season the whole irrigation system remains under water. This means that each year its canals need considerable rehabilitation work. In 1997 about 240 labour days were needed to re-excavate the *boro palla*, and 270 labour days were spent on the re-excavation of the remaining eleven *pallas*. Although in principle each household cultivating land in the command area of the system should send one male labour to participate in these works, in practice only farmers owning a substantial amount of land in the command area of the system comply with this obligation. Apparently, this 'free riding' attitude of farmers with a smaller stake in the scheme does not create any resentment.

IRRIGATION CANALS IN BEEL SINGRI

Beel Singri, the name given by the BWDB to its FCD system, erroneously suggests that it relates to only one *beel*. The area protected by this system is actually much more complex and diverse. It includes ten *beels* interconnected with each other or with the Mahananda River through a number of canals having an aggregate length of 26.5 km. These *beels* and canals are basic pillars of the livelihood system of the local population. The storage of surface water is particularly important in this region, where the annual rainfall is Bangladesh's lowest and groundwater is accessible only at unaffordable costs.

→ are used to finance various village development projects, including the maintenance of locally constructed hydraulic infrastructure.

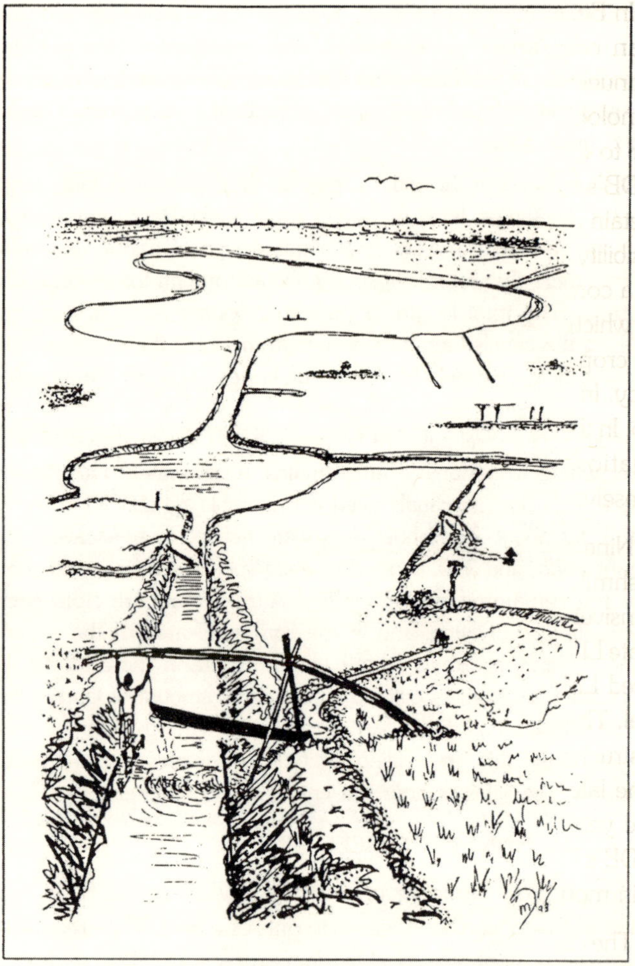

***Fig.* 5.1: Sri Narayanpur Irrigation System**

BWDB regulators have had a decisive impact on increasing human control over water. However, while Beel Singri's sluices besides providing flood control were intended to facilitate drainage, there is a large portion of the local population who would rather use them to retain water. This would allow a considerable number of farmers, most of whom cannot afford costly groundwater irrigation to cultivate winter crops such as wheat, *boro* paddy or vegetables.

In Beel Singri farmers' endeavours to construct irrigation canals began over thirty years ago with the appearance of LLPs and continued as a result of the dissemination of groundwater irrigation technologies. The irrigation canals covered by this study primarily refer to those that were triggered by the recent construction of the BWDB's FCD system. Indeed, the sluices created new opportunities to retain a large volume of water in the area's *beels* and canals. This possibility, combined with the growing role of minor irrigation devices, had a considerable impact on local cropping patterns. Vast areas of land which used to remain fallow or were cultivated only with traditional *rabi* crops such as pulses and oil seeds are now planted with *boro* paddy. In many areas the *boro* crop has replaced the traditional *aus* crop. In the context of this study we identified twelve rather impressive irrigation canals constructed and maintained by the farmers themselves.

Nine among the twelve irrigation canals are located around Beel Lakshmikol. This is one of the largest *beels* of the area and is extensively used for irrigation. The collective initiative was led by private LLP owners, who constructed nine gravity flow irrigation canals. Indeed LLP's are required to pump the water from the *beel* to the canal. The nine irrigation canals around Beel Lakshmikol have been constructed over a period of several years. The first canal dates back to the late 1960s, while the most recent one was constructed only three years ago. Seven canals were built after the construction of BWDB's Fulbari regulator in 1984, which created the opportunity to retain more water in the *beel*.

The canals around Beel Lakshmikol are fairly sophisticated pieces of infrastructure. They all are 4-5 m above ground level at the head end and 40 cm high at the tail end, having a width of 3-4 m at the head end and 1.5-2 m at the tail end; and their length varies from 70 m to 265 m. The construction cost of these irrigation canals varied between 8,000 Tk and 15,000 Tk. This represents a fairly large amount of money in the context of Bangladesh's rural economy. The high value of these canals becomes evident by calculating what a public agency would have paid to construct them. A rough calculation, using BWDB's standard schedule rates, indicates that if the canal that cost around 15,000 Tk was constructed by the BWDB, it would have cost not less than 400,000 Tk.[3] This calculation does not include administrative costs and BWDB's establishment costs.

The command area of each of these canals varies from 30 to 40 acres. Their yearly maintenance cost is about 4,000 Tk. The funds required for the construction of the canals and the contributions to their yearly maintenance are raised by the LLP owners from among all cultivators of the command area. For lands owned by absentee landlords, contributions are made by tenants. The cultivators have the choice of contributing to the construction works with labour or with cash. Many small farmers and tenants thus contribute with their own labour. Irrigation fees charged by the LLP-owners range from 600 Tk to 750 Tk per acre plus 180 kg of paddy per acre as a contribution to the pay of the pump manager and two fieldsmen. Considering the market value of paddy after harvest, irrigation under these LLP schemes costs around 1,200-1,400 Tk per acre, which, compared to the irrigation fees charged under STW and DTW schemes, is a good price.[4]

Before the construction of the Fulbari regulator, which makes it possible to retain water in the beel during the dry season, vast areas of land around Beel Lakshmikol remained fallow. This area is still known as *ghor dour* (field of horse races), as it remained submerged during the summer and was used for horse races in the winter. Only in the lower part of the *beel* was some local *boro* cultivated by using the indigenous *dhone* (locally called *jat*) for irrigation. Today's canal irrigation system allows for cultivation of one *boro* crop in 340 acres of land which prior to its construction used to remain fallow. Soon after this crop is harvested the whole area, including the irrigation canals, goes under water; and no other crop can be cultivated until the next dry season.

---

3. BWDB's standard schedule rates are applied in the context of works contracted to private construction agencies. I am grateful to my former colleague Jan Bron, senior irrigation engineer, for helping me with the calculations for this conversion exercise.
4. Under STW irrigation fees for boro are 2100-2200 Tk per acre and under DTWs 2,500-3,000 Tk per acre. In both cases, but particularly in DTW schemes the service tends to be much less reliable.

In Beel Singri we also found a locally constructed aqueduct and two types of canals known among irrigation engineers as 'irrigation canals in-fill'. These canals relate to the need to transport irrigation water to the other side of a canal. The canals in-fill constructed over Zia canal are similar in their appearance to a crossdam. They were constructed to increase the command area of a DTW and two LLP schemes. After the completion of the irrigation season these temporary structures are removed to prevent drainage congestion. They are once again constructed before the beginning of the following irrigation season. In two cases the construction work done by hired labourers employed by the owners of the irrigation scheme and this cost is included in the irrigation fee. One canal in-fill was constructed by the farmers themselves. Setting up the 35 m temporary aqueduct (made of corrugated iron sheets erected on bamboo pools) over Pagla Khari cost around 10,000 Tk, and the construction costs of the 14 m long canals' in-fill range between 2,000 and 4,000 Tk.

Case 2 — Paharpur Gravity Irrigation Canal

Mr. Humayun Chowdhury is a farmer from Paharpur. He owns 16 *bigha* (5.2 acres) of land in the proximity of Beel Lakshmikol. In the past this land could be used only as a grazing ground, for it remained submerged during the summer and could not be cultivated without irrigation during the winter season. Mr. Chowdhury decided to purchase an LLP and install it in Beel Lakshmikol about eighteen years ago. The LLP had an irrigation capacity that exceeded his own requirements; accordingly, he thought of selling water to farmers cultivating land within the command area of his pump. Given the fact that the fields are higher than the *beel* from where the water was to be pumped, before starting this irrigation scheme it was necessary to construct an irrigation canal with a slope that would allow the water to flow downward to the fields. The construction of such a canal was beyond the financial means of Mr. Chowdhury, who decided to collect the required money from the 27 farmers who would later have access to these new irrigation facilities. All of them expressed their willingness to participate in this initiative by making a financial contribution of 220 Tk for each *bigha* of land owned within the LLP's prospective command area. In this way Mr. Chowdhury collected a total of 12,000 Tk. The contract to construct the canal was given to a group of local labourers. Mr. Chowdhury personally supervised the work. Some finishing tasks were completed by the farmers of the command area on a voluntary basis. The canal requires maintenance every year. Each farmer contributes to the related costs in proportion to the land owned, on the order of 20 Tk per *bigha*. Last season eight labourers were employed for two days for this task. Mr. Chowdhury's LLP scheme employs a driver and a scheme manager who are responsible for the pump operation and ensuring an equitable distribution of the water. 250 Tk plus 65 kg of paddy per *bigha* were charged last season to each farmer to cover the cost of fuel, the rent of the LLP and the remuneration of the scheme's manager and driver. Out of each 65 kg paddy, the scheme manager received 10 kg and the driver 15 kg, while the remaining amount went to Mr. Chowdhury. All canals related to this scheme are unlined. The irrigation scheme has a command area of 80 *bighas*, and it benefits 27 farmers. It is well managed and the farmers are all very satisfied. The irrigation charge is considered fair, particularly as compared to groundwater irrigation, which would cost 800-1000 Tk per *bigha*.

Case 3 — Construction of a Canal in-fill over Zia Canal

Mr. Shafiqul Islam is the lessor and scheme manager of a DTW owned by the Barind Multipurpose Development Authority (BMDA).[5] Some years ago Mr. Islam realised that the DTW's command area could be increased if he could transport the water to the other side of a natural canal, known as 'Zia Khal.' He contacted the farmers of the potential command area and asked them if they would be interested in purchasing water from his DTW scheme, and if so, whether they were willing to participate in the construction of a canal in-fill over Zia Khal. About 30 farmers expressed interest and participated in the construction work. Some farmers sent their domestic labourers, while others contributed their own labour. Only one day was required for the construction of this structure. No material other than earth was

5. The BMDA is a remnant of the BADC, the government agency that was in charge of the dissemination of minor irrigation technologies until the sector was privatized and the agency dismantled. In the Barind area the government considered it justified to remain involved in this sector, for this dry region's agriculture development depends on DTWs, a technology that can only survive if coached and subsidized by the state (IIMI 1995).

> needed. Every year since 1990 this irrigation canal has been constructed in the month of November. At the end of the irrigation season it is removed in order to avoid drainage congestion. The canal is entirely constructed with voluntary labour. According to Mr. Islam, if the canal were constructed by wage labourers it would cost around 2,000 Tk. Thanks to the construction of this canal an additional 15 acres of land located beyond Zia Khal has been brought under irrigation during the *boro* season.

IRRIGATION CANALS IN GALACHIPA

Before Galachipa was 'polderised' by the BWDB some twenty-five years ago, it used to be a *char*, an unstable sandy island surrounded by mighty rivers. The area, in the southern coastal belt of Bangladesh, was very prone to tidal surges of saline water. After the construction of the coastal embankment around the island, life and agriculture became much safer. Nevertheless, cropping intensity and yields are still rather low, as the local conditions, particularly in relation to the availability of water, are not particularly favourable. The main crops in Galachipa are *aman* and *aus*, which, due to the often insufficient rainfall during some critical growth stages, would highly benefit from supplementary irrigation. Groundwater irrigation in Galachipa is completely absent due to the low water table; and surface water inside the polder is scarce. Tidal irrigation is the most common irrigation practice in Galachipa. It consists in taking advantage of the tides to allow the water to enter the polder through pipes or sluices installed along the embankment. This tidal water either flows to the fields through small channels or it is stored in canals, from which it is lifted with the traditional swing baskets, locally called *kechi*. If the water is not immediately used, small crossdams are built through the canals to retain the water. In some low areas, tidal irrigation can be practised throughout the month. In higher areas it is only feasible during the spring tide, that means the days close to full moon and new moon. In the winter season, when the river discharges are low, the saline water from the coast moves upstream. Accordingly, the high level of salinity makes tidal irrigation impossible in the southern part of Galachipa.

In the northern Union of Dakhua, tidal irrigation can be practised during the whole year and is extensively used during the *aus* and *aman* seasons as well as for small portions of land cultivated with winter crops, such as pulses and chilli. In the remaining areas tidal irrigation can only be practised from April to October, that means to provide supplementary irrigation to the *aman* crop. In Galachipa tidal irrigation played a somewhat more significant role prior to the construction of the BWDB embankment, when it was enough to cut small incisions in the natural river levees to allow the water to enter the land through natural canals or man-made irrigation canals. The embankment, however, has obstructed these indigenous systems and, due to the small number of pipes through the embankment and a poorly developed canal system, only about 250 acres of land were effectively irrigated with this system until recently. Tidal irrigation was revived at the beginning of the 1990s, within the framework of a water development project funded by the Government of the Netherlands. Through a participatory approach, farmers selected the sites for the construction of 24 new irrigation inlets. They were further motivated to provide their own resources to develop a canal system that would optimise the use of scarce water and maximise of the command area of each irrigation inlet. The farmers of these areas donated their land and constructed the new irrigation canals with their own means, either by providing labour or with financial contributions proportional to the amount of land owned in the command area of an inlet. This initiative, in which external agents played an important role, materialised entirely through the mobilisation of local resources; and it brought an additional 800 acres under irrigation. These tidal irrigation schemes are used for supplementary irrigation of *aus* and *aman* as well as for some *rabi* crops.

Crossdams

Crossdams are common indigenous water management structures that can be found all over Bangladesh. In most regions they are called *badh*, except in the south-east, where they are known as *goda*. Crossdams are earthen dams, generally reinforced with bamboo poles, banana stems and bamboo mats; they are built across canals or streams. Some are small and can be constructed by a few labourers within one or two days. Others are so large that construction may involve several thousand labour days. The most common function of crossdams is to retain water used for distinct purposes such as irrigation, fishery or domestic uses. Several crossdams have the function of preventing the intrusion of water into a certain area. The same crossdam may simultaneously fulfil more than one function or it may change its purpose in distinct seasons. In Chaptir Haor, Beel Singri and Galachipa crossdams are constructed to maximise the retention of surface water in *beel*s and canals. This water is primarily used for agriculture but in some cases — particularly in Beel Singri and Galachipa — it also meets important domestic needs. Many crossdams found in small canals in the interior areas of Galachipa were built by private individuals to create ponds now used for fish culture.

Two crossdams in Galachipa were constructed in the middle of canals which are closed at both ends with sluices, with the objective of creating two independent hydrological units, that is two areas within which it is possible to control water levels through the operation of BWDB's regulator. A very large crossdam was constructed almost thirty years ago to protect the area from tidal surges. Among the 35 crossdams found in Chaptir Haor, only two retain water exclusively for domestic uses such as bathing and washing cattle or kitchen utensils. The remaining 33 crossdams are used either for both domestic and irrigation purposes or exclusively for irrigation.

Most crossdams are temporary structures that are re-constructed every year. They are generally built at the end of the monsoon, after the excessive water that accumulates during the rainy months has been drained. They need to be removed at the end of the dry season to prevent drainage congestion. Removal is not usually a big job, for crossdams naturally wash away or are damaged by the passage of boats. The construction of crossdams is often ensured by groups of farmers through voluntary labour. In several cases, particularly in Galachipa and Banskhali, their construction is entirely financed by private individuals who thereby appropriate the exclusive right over the dammed water. In Chaptir Haor, crossdams either serve specific groups of farmers or have a more public function. This has clear implications for their provision. In the first case their construction is ensured by contributions in labour or cash from direct beneficiaries, whereas in the second case the village may finance construction in part with locally mobilized *panchayat* funds. For example, one village in Chaptir Haor allocates each year 6,000 to 7,000 Taka of its common fund for the construction of 18 crossdams. Since this fund is insufficient to finance the whole work, 40 or 50 people participate in their construction on a voluntary basis. Some of Chaptir Haor's crossdams primarily benefit one or two large farmers. They take upon themselves the full construction costs even though they may allow other people to use the retained water. This is the case for two dams built by a large farmer across a *mara nadi* ('dead river'), at a total cost of around 5,000 Tk. The water retained by these crossdams serves to irrigate around 150 acres of land, of which 100 acres belong to the farmer who finances this project. The water also is used for domestic purposes by the whole village population.

Case 4 — Teliakhari Crossdam in Sanpur

The initiative to construct a crossdam on Teliakhari, a major canal in Beel Singri, has been led for several years by Mr. Aftab Uddin, a village *matabar* of Sanpur. Several farmers owning LLPs and land near the *khal* support its construction, for it provides them with the opportunity to store sufficient water to irrigate their wheat, sugar cane, and some *boro* paddy cultivated in an area of about 150 acres. The water stored in the small reservoir created through this crossdam is also used for bathing, washing clothes, cooking and other domestic purposes and is much appreciated by the whole population of Sanpur. The Teliakhari crossdam is entirely constructed by voluntary labourers, with rich farmers contributing the services of their domestic labourers and poorer farmers their own labour. About 40 people are required for two days to construct this crossdam. The required materials, such as bamboo, banana stems, and bamboo mats, are donated by some villagers. Over 1,400 households living in the surrounding villages are making use of the water retained for domestic purposes, thanks to this crossdam. However, from March to May, due to excessive withdrawal of water for irrigation, the canal dries up completely. Until recently the fish in the reservoir used to be an open access resource. However, two years ago the *matabars* of Sanpur and the nearby village of Meghdohor decided to lease out fishing rights in the reservoir created through the crossdam, to generate a fund to finance the construction of a mosque, which gave to this crossdam an additional function.

According to the key actors involved with its construction, at present there are no conflicts in relation to the Teliakhari crossdam. However, some years ago the influential people of Sanpur managed to persuade the Union Parishad to support this initiative by allocating seven tons of FFW wheat. These external resources made it possible to construct a much larger dam, which retained a larger volume of water. For the people upstream of Teliakhari canal the crossdam turned out to have a negative impact; due to the high water level, they could no longer cross the canal with the bullock carts extensively used in this area for the transport of paddy and other products. The affected people became so upset, that they cut the dam in order to release the excess water. After this incident the crossdam was re-built according to the previous, more modest design. Besides this overt conflict, which occurred only once, there is a conflict of interest between the poor people, who would like to use the stored water for domestic purposes and the LLP owners and farmers who use the water for irrigation. If the water stored in this reservoir was used only for domestic purposes, it would not dry up completely in the season when people suffer a water shortage. However, most poor people who favour this option are poor sharecroppers or landless labourers who depend for their survival on land owned or employment provided by the wealthy LLP owners. They are not in a position, therefore, to defend their interests.

Case 5 — Crossdam on Bowalia Khal in Manikchand

Before Galachipa was developed into a polder by the BWDB, the area used to be severely affected by tidal surges and the intrusion of saline water. These tidal surges came through a canal named Bowalia Khal. In those days, due to this problem, much land remained fallow or could be planted with only one very low-yielding variety of *aman* which tolerates saline conditions. About thirty years ago Mr. Mozammul Hoque, who ever since that time has been the elected UP (Union Parishad) chairman of the Ratandi-Taltali, took the initiative to build a big crossdam on Bowalia Khal, thus protecting a major portion of land from tidal surges. Mr. Hoque envisioned that the construction of this crossdam, whose cost he estimated to be around 800,000 Taka, would improve about 4,000 acres of land owned by over 1000 households by allowing for cultivation of an additional *aus* crop and also some *rabi* crops. Mr. Hoque organised a number of meetings with local farmers to explain his project. They assured him that they would contribute to this initiative but they were concerned that the costs involved were beyond their means. Based on these considerations, Mr. Hoque submitted his project to the Thana Circle Officer (Development)[6] and to the Deputy Commissioner in Barisal and informed them that about 50 per cent of its costs would be borne by the local farmers. Finally, the local government agreed to contribute 400,000 Tk to the construction of the crossdam. At this stage Mr. Hoque organised five committees to collect the money from farmers owning land within the area to benefit from the dam. Each farmer was expected to contribute 100 Tk per acre of land owned. Construction of the crossdam was executed by the Union Parishad through a contractor and monitored by the Thana Parishad. Even today this crossdam has a vital function. It is part of the main road inside the Polder and connects its interior areas with the Thana headquarters. The crossdam is now periodically rehabilitated and maintained by the LGED.

ROLE OF CROSSDAMS IN BANSKHALI

The portion of Banskhali Upazila protected by BWDB's Polder 64/1b consists of a coastal strip of land located between the eastern hilly area and the Banskhali canal. It includes 36 villages inhabited by around 255,000 people (BBS 1991). The area is separated from the Bay of Bengal by a narrow strip of embanked land, belonging to the same Upazila, known to the BWDB as 'Polder 64/1a'. Banskhali is a heavily populated area. Its population density of 3290 persons per km^2 reveals an almost peri-urban character, explained by the area's proximity to Chittagong, Bangladesh's most important port city. Nevertheless, even though many local people take advantage of the region's numerous off-farm employment opportunities, agriculture continues to play a pivotal role in the local economy. The FCD scheme was constructed in 1967 with the objective of enhancing

6. This office relates to the government administration of former East-Pakistan and no longer exists today. Mr. Haque regretted that this type of government participation to local initiatives would no longer be possible in these days, because of a much more centralized government structure and because the development of local infrastructure is entirely controlled by technocratic agencies. On this subject see Barenstein (1994a and 2000a).

agricultural production by protecting the land from tidal surges and saline water intrusions. Before its construction, most land was single cropped with a local, low-yielding variety of *jali aman*. The scheme was an important catalyst for the adoption of HYV varieties of transplanted *aman* and for the cultivation of an additional crop in most land. Where water for irrigation is scarce this is bound to be an *aus* crop. Various indigenous as well as modern irrigation technologies can be found in this area (artesian wells, traditional surface irrigation systems, LLPs, STWs and DTWs), and accordingly much land is cultivated with *boro*. About 15 per cent of the land is triple cropped, generally with two paddy crops and one crop of vegetables. The commercial production of vegetables is particularly important in this area, which supplies fresh products to the urban markets of Chittagong.

BWDB's hydraulic infrastructure consists of a 53 km embankment along Banskhali Khal, a branch of the Sangu River flowing into the Bay of Bengal at the southern end of the scheme. The embankment is equipped with 32 drainage sluices to allow natural streams (locally called *chara*) flowing through the area to Banskhali Khal, and five surface sluices facilitating the drainage of pockets of lowland. Most of these sluices are in very poor condition, because of neglected maintenance and people's attempts to use them for purposes for which they were not designed. Indeed, the sluices, though provided with slots which would allow the insertion of fall-boards, were originally designed for automatic drainage and were accordingly only equipped with flap gates. This design does not take into account local water demands, which require control over the water flowing from and to Banskhali Khal.

Banskhali's crossdams are an endogenous response to this necessity. In the south-eastern hilly area of Bangladesh there are different types of crossdams. In local dialect, however, they are all named *goda*. One type is built through natural streams of spring water to create small reservoirs of water to be used for irrigation. This constitutes an ancient practice.[7] One of these ancient irrigation

7. A detailed account of these indigenous irrigation systems is given in the Chittagong District Gazetteers published by the Government of Bangladesh in 1975 (Rizvi 1975).

schemes has recently been upgraded by the LGED,[8] within the framework of a minor irrigation development project funded by IFAD. This represents a rare case of building modern irrigation schemes upon pre-existing indigenous systems. Among the 22 crossdams found in Banskhali, 16 are located in the hilly interior area, where they have been reconstructed every year since ancient times. Their cost depends on size and construction material and varies between 5000 Tk and 30,000 Tk. Although their main function is to dam water for irrigation, the water reservoirs are also used for fish culture, and the structures allow pedestrians to cross the streams on foot. Six crossdams were constructed near dilapidated sluices to prevent the intrusion of saline water and retention of sweet water for irrigation and/or fishery. These provide an interesting example of how indigenous knowledge and technologies may be used to solve 'modern' problems or to take advantage of newly created opportunities.

Irrigation is the main objective behind the construction of most crossdams. Without irrigation, it would be possible to cultivate only rain-fed *aman*. Though *godas* are to be considered an indigenous water management technology, their number and size increased significantly after the construction of the FCD scheme which protects the area from the intrusion of saline water and hence gave farmers an incentive to intensify crop production. Other factors that contributed to the increased importance of *godas* were the advent of LLPs, HYV paddy, and the growing market demand for vegetables. These factors make the retention of surface water more valuable than ever, which indicates that the advent of modern technologies rather than making traditional ones superfluous, may actually enhance their importance. *Boro* paddy is cultivated near canals by lifting the retained water by means of LLPs or in areas covered by groundwater irrigation. The

8. The Local Government Engineering Department is an important government agency involved in rural infrastructure development. Initially the LGED was primarily active in the construction of rural roads but gradually — thanks to its privileged status among international funding agencies impressed by its charismatic leadership and comparatively entrepreneurial management — it also 'encroached', to their great dismay, into water management, a domain that was originally controlled by other public agencies, especially the Department of Public Health Engineering (DPHE) and the BWDB.

smaller vegetable fields and seed beds are irrigated with the help of pitchers or swing baskets, locally called *hichuni*, and also in areas covered by groundwater irrigation. All people involved in the construction of crossdams were quick to suggest that the government should re-excavate the silted canals as to allow them to store more water for irrigation. No such initiative has been taken locally, as the related cost probably exceeds people's perceived direct benefit.

Another 'modern' feature of Banskhali's traditional *goda* relates to prevailing provision patterns. Those crossdams located near BWDB sluices, for example, are constructed by individual or informal groups of 'entrepreneurs', who invest in their construction with the aim of selling water. Two examples describing typical Banskhali crossdams and their related organisational practices are given below.

Case 6 — Construction of a Crossdam on Jaldi Chara

Jaldi Chara is a natural stream (*chara*) flowing down from the Eastern Hill Tracts through a village named Jaldi. A group of farmers from Jaldi constructed a dam across this stream for the first time over thirty years ago with the aim of storing water for irrigation. In those days the water stored in the small reservoir created by the dam was distributed to surrounding fields by means of traditional irrigation devices. These days, however, its yearly construction is entirely managed by the owner of an LLP. The labour costs involved in the construction of this earthen dam, which is 7 m long, 3.5 m and 1.5 m wide, is about 6000 Tk. The LLP owner collects the required money from farmers owning land within the scheme's command area at a rate of 150 Tk per acre. Three sharecroppers contribute to the construction of the crossdams with their labour instead of money. The dam allows for storage of a volume of water sufficient to irrigate 30 acres of land. As may be expected in an area where over 85 per cent of the farms cultivate less than 2.5 acres of land,[9] the scheme's command area belongs to 48 different farmers, who need water for their *boro* paddy or winter season vegetables as well as for their summer *aman* crop. The LLP owner charges the farmers an irrigation fee of 1500 Tk per acre. This irrigation scheme has increased the yields of the *aman* paddy and has made it possible

9. GOB/BBS (1987).

to plant a crop in the winter season. The farmers cultivating land within its command area are understandably quite satisfied. However, the crossdam creates resentment among farmers downstream along Jaldi Chara, particularly in periods of drought. In critical periods, to protect the dam from sabotage, the LLP owner has employed a night guard.

Case 7 — Construction of a Crossdam near a BWDB Sluice

Several of Banskhali's crossdams are built by local people near dilapidated regulators to prevent the penetration of saline water and/or leakage of sweet water. This is the case of the crossdam constructed on Kodala Chara near a dilapidated BWDB sluice. The initiative to construct a dam across this stream in the village of Chapachari was taken for the first time in 1992 by an informal committee of seven people, as a response to the deplorable condition of the sluice. Indeed this structure is currently useless, for its six vents are all missing. The crossdam aims at preventing the intrusion of saline water and retaining the sweet water, carried by the natural stream, for irrigation. It is constructed each year in the month of October and removed in May, after the harvest of the *boro* crop. The local BWDB office is informed about this practice through the committee, who sought their permission prior to investing in its construction. The dam costs around 12,000 Tk, a sum advanced by the seven initiators. They then charge a fee of 100 Tk per acre from all farmers who want to use the retained water for irrigation. Though the concerned farmers pay this money, they are openly grumbling about the fact that the seven members of this informal committee claim the exclusive right to fish in the reservoir. Given the profitability of this secondary function of the dam, they argue, the committee should not charge any fee from the farmers.

'Janghals'

Janghals constitute an ancient water management technology with multiple functions unique to the *haor* basin. First of all, *janghals* serve the purpose of what are known among hydraulic engineers as 'contour bundhs,' that means dikes built along contour lines to retain water and moisture within a certain area to reduce the need for irrigation. They are a well-known technology in other countries, for example, in the Sahel region but nobody ever appears to have noticed that they exist in Bangladesh.[10] Sec-ondly, many *janghals* have a vital role in providing subsidiary flood protection. This means that if the main embankment should fail, *janghals* may save crops from flood by protecting the land within their command areas. As will be illustrated through a specific case later in this chapter, thanks to a *janghal* that in 1996, when an embankment breached, it was possible to save almost the entire crop in the *haor*. Thirdly, in the dry season *janghals* serve as roads to transport paddy to homesteads or to threshing grounds. In some cases *janghals* provide the only road links between villages. This auxiliary function is important because in the *haor* the harvest has to be completed within the shortest possible time, that is, before the area is submerged. One *janghal*, locally known as *Chosmer janghal,* shortens the walking distance between the *upazila* headquarters and ten villages by about 2 km and is extensively used by the pedestrians. However, it is perhaps because *janghals* are so extensively used as roads that outsiders never realised that their primary function was a hydraulic one.

In Chaptir Haor there are 92 *janghals* with a total length of more than 50 km. Each *janghal* has a length varying between 250 m and 4 km, a height of 1-1.5 m and a width at the crest level of 1-3 m. Nobody in Chaptir Haor is able to recall when and by whom their *janghals* were originally constructed. However, each year people invest considerable human and material resources in repairing and maintaining them. Indeed, the preservation of *janghals* is one of the main public concerns in Chaptir Haor and constitutes the principal impetus for organised collective action. Every year, at the end of the monsoon, after having been completely submerged for about six months, *janghals* need to be rehabilitated. In the case of small *janghals* these works are carried out by their direct beneficiaries, who contribute either with their own voluntary labour or, in the case of large farmers, by sending their domestic labourers. In some cases funds are collected among beneficiaries and the task of repairing

---

10. The only scholar who has noticed the existence of this infrastructure is Islam (1985). However, all the mentions on the subject are that *janghals* or internal embankments are constructed for drainage and irrigation purposes' (Islam 1985: 31).

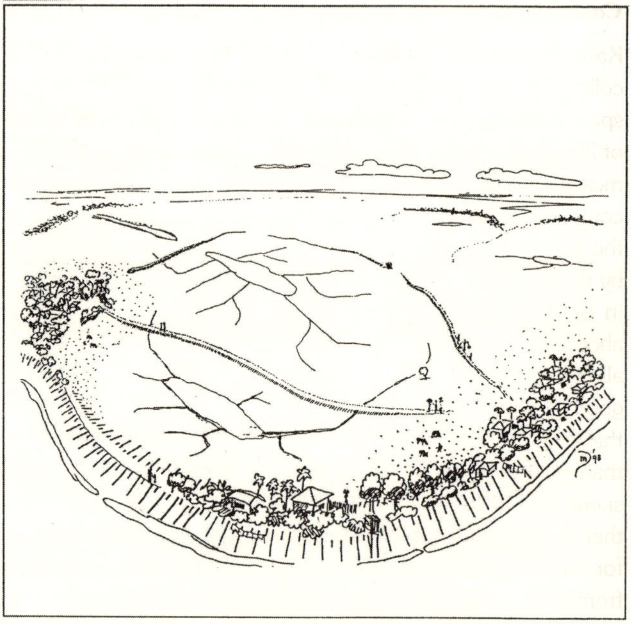

***Fig.* 5.2: Chaptir Haor's Janghals**

janghals is delegated to gangs of local labourers on a fixed contract basis. In these cases, poor farmers have the option of contributing their own labour instead of cash. Large *janghals* — in particular those benefiting one or more villages — are repaired by labourers employed on a contract basis by the *matabars*. The money to fund these works is drawn from *panchayat* funds.[11] The volume of earth work that needs to be carried out each year to maintain each *janghal* is fairly predictable, and accordingly some villages make sure to save for the time when monetary contributions are needed. For example, the village of Karimpur, where there are four *janghals* with a total length of approximately 2300 m, allocates each year 40,000 to 45,000 Tk of its *panchayat* funds for the annual *janghal* maintenance works. This fund pays wages for some 900 labour-days. Around 8,000-10,000 labour days are required each year to maintain all of Chaptir Haor's *janghal* systems.

11. Details about the *panchayat* funds are given in Chapter 6.

Case 8 — Karimpur's Janghals and Palla

Karimpur, a village located in the north-west of Chaptir Haor, collectively owns four *janghals*. The local people know them by specific names, such as *Jhur beeler janghal* (the contour bundh of Jhur beel), *Nayer bonder janghal* (contour bundh of the Nayer meadow,[12] *Medar bonder janghal* (contour bundh of the Medar command area), and *Chosmer bonder jangha* (contour bundh of the Chasma command area). These *janghals* were constructed by the ancestors of Karimpur's and Matiapur's current inhabitants in order to retain moisture and irrigation water in an area of about 200 acres. During the dry season, one *janghal* is used by all villagers as a road and has a pivotal role for the transport of goods and people. Every year, after the monsoon water recedes, the *janghal* needs considerable repair and maintenance. Rather than ensuring this through voluntary labour, the people of Karimpur spend each year 40,000 to 45,000 Tk for the maintenance of their *janghals*. About 1,000 labour days were deployed generally for the maintenance of these four *janghals*. This money is drawn from Karimpur's *panchayat* fund, a fund generated through leasing out *dhoba* and through taxes collected during village fairs (*melas*), which are primarily organised for this purpose. Within the command area of Karimpur's four *janghals* there are 30 large irrigation canals and 50-60 smaller ones. These also need annual maintenance, which is ensured by voluntary labour. Each household owning or cultivating land inside their command areas —irrespective of the amount of land owned — is expected to send one labourer to participate in such yearly maintenance work at a day fixed by the farmers' informal leaders.

Cutting and Closing Embankments

Embankment cuts are a well-known practice, generally referred as 'public cuts'. From the BWDB's perspective, these 'public cuts'

---

12. *Bond* is a local term referring to the a large meadow. In relation to irrigation, however, the term *bond* basically refers to the command area of a certain *beel*, *dhoba* or irrigation canal. *Janghals* have the function to retain the water in these command areas. In the *haor* plots are not delimited through bundhs. Large fields may be owned by several people, while the plots owned by a particular person are either demarcated by stones or by *sangha*, a flood resistant grassy plant commonly used in low land areas for demarcating plots.

represent a criminal offence and the people behind them can be legally prosecuted. Indeed, the BWDB often files legal cases against the 'unknown public' responsible for what many engineers perceive as sabotage. From another perspective, as was discussed in the previous chapter, environmentalists and social activists interpret 'public cuts' as the ultimate proof that the people are against embankments. For international development agencies and their experts 'public cuts' are the response to a hydrological problem that could otherwise be solved through the construction of sophisticated regulators. This technical solution to public cuts, however, has the disadvantage of making people dependent on infrastructure that is unaffordable without international aid and which has high annual maintenance costs. Cutting and closing an embankment is less expensive. In fact, this study has revealed that embankment cuts are common and in many cases practical and cost effective water management practice, whose technical and economic viability would be worth considering before making any area dependent on external agencies.

Embankments are generally cut in a well-planned and systematic way; and unless people have the certainty that someone else (e.g., BWDB, LGED, Union Parishad) will close them again, they make sure to do it themselves. In the context of this study, we identified 20 embankment cuts, out of which 15 are routinely closed again by the people themselves. In the remaining cases this responsibility has been taken over by the BWDB and, in one case, by the LGED.

The highest number of embankment cuts was found in Chaptir Haor, where they represent a traditional drainage practice that both facilitates final fishing and navigation and creates the conditions necessary to prepare the land for planting *boro* paddy, the only crop that can be grown in the *haor*. There is a formal agreement between the farmers and the leaseholders that no embankment will be cut before *Phalgun* 15th (beginning of March). In any case, the exact date of the cut is always discussed in advance with representatives of the farmers.

In Banskhali embankment cuts aim to relieve some low land areas from drainage congestion causing the damage of the *aus* crop, hindering communication and submerging homesteads and ponds.

The cuts are made up to three or four times in a season and immediately closed again. The cuts are always done at the same places where people have installed drainage pipes. Under normal conditions, these pipes are sufficient to release the area from drainage congestion.

The cost and labour involved in cutting embankments is minor. However, considerable labour is required to repair them. In Chaptir Haor, the leaseholders send their seasonal labourers to close these cuts. In Banskhali the benefited people participate with a cash amount varying between 30-40 Tk per acre of land owned within the area benefited by the cuts.

The concern shown by the people of Banskhali and Chaptir Haor to fill in the cuts immediately after having released congestion, shows that embankment cuts should not always be interpreted as a sign that people do not appreciate embankments. Even if embankments cause drainage congestion during specific periods of the year, they still may be considered vital during other periods.

The only two 'public cuts' with a history suggesting overt tensions between the local population and the BWDB were found in Galachipa. In those cases the people who were affected by severe water-logging repeatedly appealed to the BWDB to take action. Only when their pleas remained unanswered did they decide to solve the problem with their own means. In one case it has become a routine that the embankment is cut each year by the people and then closed again by the LGED. In the second case the 'public cut' consisted in removing a temporary crossdam constructed by a BWDB contractor.

THE LOGIC OF 'PUBLIC CUTS' IN CHAPTIR HAOR

Chaptir Haor is encompassed by about 43 km of natural river levees. The local population have long understood that by raising these levees they could provide better protection to their crops. In the early 1980s the BWDB further raised these embankments. With its intervention the BWDB responded to the *haor*'s need for flood protection, but the agency failed to consider the pivotal importance of drainage.

While the people of Chaptir Haor basically welcome this improvement, protecting their crops from flood is an important but only temporary concern. At the end of the monsoon, cultivators' main preoccupation is to drain their fields so that they can start to prepare their seedlings and plough their lands. Equally interested in drainage are the *beel* leaseholders, for whom drainage represents an effective fishing technique.

Drainage is effectuated through 'cuts' in the embankment at the point where natural drainage canals, locally named *khara,* meet with the embankment. Under natural conditions, that means that if there were no embankment the canal would constitute a direct link between the rivers and the *beels*. Such natural conditions, however, did not exist prior to BWDB's intervention, as the people themselves — before the beginning of the monsoon — would build dams to prevent river water from entering into the *haor*. If they were not washed away naturally or destroyed by the passage of boats, people removed them at the end of each monsoon season.

Chaptir Haor's embankment is cut in thirteen different locations. In two locations the cuts are exclusively made by farmers with the objective of removing drainage congestion in low lands in order that they might start planting paddy. In one location boatmen cut the embankment in summer to facilitate water transport of people and goods. In the remaining ten cases the initiative to cut the embankment is taken by *beel* leaseholders. Cutting an embankment is not a one-time activity. It is done gradually, starting from the Bengali month of *Agrahyan* (November-December). Since the water in the *beel* is a common resource used for irrigation, before cutting the embankment leaseholders have to seek permission from the farmers whose land is located within the concerned *beel*.

The approval to cut the embankment is given by village *matabars*. In the context of formal meetings they also reach an agreement on the water level to be retained in the *beels*. There is an institutionalised agreement between farmers and leaseholders that final fishing through complete drainage is not allowed before *Phalgun* 15th or the beginning of March. In any case, the exact date always has to be agreed upon by the farmers whose interests are represented by the *matabars*. *Matabars*, besides being expected to guard the village interests, are generally themselves large farmers and have a keen interest in making sure that enough water is retained inside the *haor* to allow for irrigation.

Traditionally, in those cases where cutting the embankment was aimed at fishing, the costs involved in cutting and closing the embankment were borne by *beel* leaseholders. They had to deposit an agreed upon amount of money with village *matabars* in the context of a public meeting, as a sort of insurance that they would again close the cuts after having completed the final fishing. Nowadays several cuts are closed by the BWDB under FFW programmes. Through these interventions, not only are public resources spent to the benefit of a few privileged people but local accountability has been significantly reduced. The leaseholders highly benefit from their exclusive control over common resources and certainly have the means to pay for refilling the embankments, which they themselves cut.

The *haor* people, of course, are fully aware that cutting an embankment is a risky activity. The main public concern in the case of this indigenous water management practice, however, is not the mobilisation of labour and material resources. Rather it is ensuring through informal rules that the cuts are done in conformity with local needs.

Case 9 — Embankment Cut at Kalibhui Khara

Every year, at the beginning of the Bengali month of *Kartik* (October-November), the leaseholder of Kalibui Beel sends two labourers to cut the embankment at Kalibui Khara (Stream) and install his gear to catch the fish carried by the current. The labourers work on this spot for about 15 days. Each day they spend about three hours to deepen it. Although the primary function of the cut is to fish, farmers support this drainage practice, which they need to prepare their fields for paddy cultivation. About 1,400 acres of land owned by more than 1,000 households are benefited from cutting this embankment. However, if farmers notice that an excessive amount of water is drained, they order the leaseholders to close the cuts immediately. Final fishing, which involves the complete drainage of Kalibui Beel, is never allowed before *Phalgun* 15th (beginning of March). To fix the exact date, the leaseholder meets with the *matabars* of

Matergoan and Barakapanpur. After cutting the embankment for final fishing, the breach is closed by the BWDB. However, before Chaptir Haor was taken over as a project by the BWDB, the leaseholders were responsible for closing the breach after final fishing. To this aim, they had to submit in advance an agreed upon amount of money corresponding to the cost involved in closing the embankment. This money was handed over to the *matabars* in the context of a public meeting.

Case 10 — Removal of a Crossdam near the BWDB Regulator on Bowalia Khal

Bowalia Khal represents the main drainage canal of Galachipa and discharges most water accumulated in the surrounding areas. In July 1994 two crossdams were built by a BWDB contractor rehabilitating the regulator on Bowalia Khal on the land side close to the Bowalia regulator. The construction of these crossdams caused severe water logging in about 1000 acres of land. Some large and influential farmers of the area, together with the UP members and chairman, personally visited the BWDB office and requested its immediate action to solve the problem. The BWDB, however, did not pay any attention. As a result, the upset people organised a big demonstration and with the backing of the Union Parishad and the TNO cut both crossdams. About 5,000 people gathered at Bowalia Khal, though the work involved in cutting the crossdam only engaged ten voluntary labourers for about one hour. This local initiative released 1000 acres of agricultural land from drainage congestion and benefited about 5,000 households.

Case 11 — Cutting and Closing the Embankment at Atrai Khara

Every year in the Bengali month of *Phalgun* (February-March) the leaseholder of Atrai Beel seeks permission from the farmers of three villages whose lands are located near Atrai Beel, to cut the embankment at Atrai Khara. This drainage practice usually takes place at the end of the fishing season. Fishing gear and nets are fixed at the location of the cut to catch the fish carried away by the current. The leaseholder starts by making a small incision in the embankment, and then gradually enlarges it into a big cut as the water level in the *beel* decreases. The farmers agree with this cut, which they also need to start ploughing the lower lands around the *beel*. In fact, the cultivation has to start as early as possible due to the increasing risks of hail storms and flash floods towards the end of the season.[13] Nevertheless, the farmers closely monitor the water level in Atrai Beel, as they are equally concerned with retaining sufficient water for irrigation. Whenever the desired water level is reached, they instruct the leaseholder to close the cut immediately. The leaseholder, however, is not overly concerned to fill in the cut very thoroughly and the farmers need to strengthen the closure.

About 20-25 farmers owning land in the command area of the *beel* are involved each year in strengthening this closure. There is no formal rule on how much labour each farmer in the command area has to provide for this purpose. Only labour and earth, which are easily available on the spot, are required for this work. Under normal conditions the embankment closure is done after completed drainage is sufficiently strong to resist the pressure of the rising river in the month of *Baishak* (April-May). However, the spot where the cutting and closing are practised is known to be vulnerable. It needs to be watched and, in case of extraordinary water levels, raised and strengthened. A breach of the embankment at Atrai Khara not only would affect the farmers owning land in the immediate surroundings of the *beel* but also would threaten the crops in some 2,400 acres of land belonging to over 700 households.

The need to raise and strengthen the closure at Atrai Khara in relation to early floods occurs about every second year. The risk of the closure at Atrai Khara being overtopped or breached is considered an emergency of public concern and all people from the surrounding villages are urged by microphone to participate in the necessary works. The last time such an emergency occurred was in 1996. About 150 people from five villages worked for a

13. The need to harvest the paddy as early as possible is a major constraint for the adoption of HYV. In fact, the maturation period of local paddy varieties is 120 days, whereas HYVs require 140 days. This means that while the harvesting of local varieties can begin in the Bengali month of *Chaitra* and be completed by mid-*Baishak*, the harvest of HYV begins during the first week of *Baishak* and is only completed by the first week of *Jayshta*. The risk of hailstorms is particularly high in *Baishak*, and the risk of floods gradually increases during the same month. Accordingly, HYV crops are only cultivated on small portions of relatively high land.

> full day to raise and strengthen the Atrai Khara closure with bamboo, bamboo mats and sand bags. The works continued with a reduced number of people for another three days. All work was voluntary and the required materials were donated by wealthy farmers.

Drainage Outlets and Irrigation Inlets

Drainage outlets and irrigation inlets consist of pipes installed through an embankment or through the bank of a *khal*. The installation of such pipes, called *'nasi'* in the south-eastern dialect (a word derived from 'nose'), is fairly common in coastal areas and was found in almost equal numbers in Galachipa and Banskhali. The same pipe may serve a number of purposes, such as releasing water-logged areas from drainage congestion, tidal irrigation, fishing and water management for salt and shrimp production.

There were in all 25 pipes found in Galachipa, located in Taltali Khal, a canal located in the north-east of the polder. After the canal was re-excavated by the BWDB five years ago, the farmers realised that it would now be possible to use tidal water for irrigation by placing pipes through its banks. Through each of them it is possible to provide supplementary irrigation to *aus* planted in a command area of 3-6 acres. The farmers received the pipes free of charge from the BWDB but all costs related to their transport and installation were borne by them.

Most of the 23 pipes installed by individuals or groups of people in Banskhali serve primarily to release pockets of low land from drainage congestion. They may at the same time be instrumental for fishing, and in a different period of the year, be used for tidal irrigation. BWDB hydraulic infrastructure in Banskhali is highly appreciated by the local population, for it has made a substantial contribution to agricultural production by protecting the land from tidal surges. However, due to the uneven topographic features of the land in that area, there are a large number of lowland pockets, which as a result of the embankment now are very prone to water-logging. A total of 23 pipes have been installed over the last 20 years by the people of Banskhali. Among them, 17 serve mainly to release pockets of low agricultural land from drainage congestion; but in some cases they are equally often used for tidal irrigation. These pipes also create an opportunity to catch fish. Six pipes have been installed for the purpose of salt production.

Banskhali's pipes usually are made of cement. They have a diameter of 15-45 cm and a length of 18 m. They can be purchased in local markets from the Department of Public Health Engineering (DPHE) or from a national NGO which produces them as part of an employment generation project for the poor. The cost of a cement pipe ranges from 3500 to 6000 Tk. Before the widespread availability of cement pipes, people used hollow coconut or palm trees. These had the disadvantage that they needed to be replaced after about one year. They can still be found in more remote areas, around Teknaf, for example. The command area of one pipe ranges between 3-20 acres, an amount of land that is generally owned by more than one household. It is often the case, however, that very few households are acutely affected by drainage congestion. Accordingly, it was found that the expenses related to the installation of 12 drainage pipes which are entirely borne by individual households, although several other households benefited from these installations. In such cases, those who take upon themselves the burden of installing the drainage pipe have the exclusive right to catch fish by installing their fishing gear at the pipe's opening. Whenever the installation of pipes was financed by collective resources, financial contributions of the order of 50-100 Tk per acre of land were collected from among the landlords who get the benefits by the initiative.

Drainage pipes need regular maintenance, for they get easily clogged with silt and plants. The responsibility for maintaining the pipes generally is given to the family living nearest to the site where the pipe has been installed. In some cases, as compensation for their labour, they obtain exclusive fishing rights at the drainage outlet. Considering that during the monsoon it is possible to catch up to 3 kg of fish per day, this is a fairly good enough compensation. The capacity of these pipes is not always sufficient to reduce severe drainage congestion resulting from strong rainfalls or flash floods from the hills. In these situations the affected people need to cut the embankment.

From people's point of view, the installation of drainage pipes is a necessary and cost-effective response to drainage congestion. It also represents a water management practice around which there are hardly any conflicts. However, the installation of pipes sometimes damages and weakens the embankment and accordingly threatens the safety of the whole flood protected area. These long-term effects are often not fully understood or taken into account by the people, who are more concerned with solving their immediate problems. This shows that even if people are active and willing to pay for improved water management, full responsibility cannot just be turned over to them. The solution is neither to forbid nor to ignore these initiatives nor to replace relatively cost-effective and appropriate initiatives with expensive and externally financed infrastructure. The BWDB could play an important role in monitoring, advising and assisting people in their initiatives and ensuring that these do not threaten the viability and sustainability of public hydraulic infrastructure.

Case 12 — Installation of a Drainage Pipe in Napura

This initiative was triggered by a severe drainage congestion problem that affected about 75 acres of land owned by fifty families from the village of Napura in Banskhali. The drainage congestion was a recurrent problem from June to September, when the area is subject to flash floods coming from the nearby hills. These flash floods used to damage *aus* crops, homesteads and ponds. For a number of years the affected people solved their problem by cutting the embankment, which they would close again after releasing the water, for they were equally concerned to protect the area from the intrusion of saline water. About three years ago a primary school teacher, who was also known in the village as an innovative farmer, motivated the villagers to find a permanent solution to this recurrent problem. A cement pipe with a length of 6 m and a diameter of 45 cm was purchased in the local market for 4,500 Tk. About thirty voluntary labourers and 20 labourers paid by wealthy farmers worked for one day. The costs related to the purchase and the installation of this drainage pipe were divided among the fifty families in proportion to the land owned, at a rate of 60 Tk per acre. The maintenance of the pipe was assigned to the owner of the land where the pipe was installed. As compensation he was given the exclusive right to fish at the location of the pipe outlet. An informal committee was formed for the purpose of managing this pipe. Under normal conditions this committee is inactive, but in case of an emergency its twelve members are expected to take action. The people of Napura were all satisfied with this initiative, which has not had a negative impact on any group of people. Under normal conditions the pipe has the capacity to relieve the area of drainage congestion. Only in case of severe flash floods it is still necessary to cut the embankment. Whether and when to take such action is decided by the committee.

Raising Roads and Embankments for Flood Protection

Collective actions to raise embankments and roads to protect homesteads and agricultural land from flooding are relatively common. Twelve such initiatives were identified in Chaptir Haor, three in Beel Singri and one each in Banskhali and Galachipa. These initiatives generally occur in a moment of crisis, that is, when a specific area is threatened by flood. In these situations there is neither time to apply for external assistance (although in several cases people have appealed to the BWDB or other agencies for support) nor to implement strict rules and financing modalities. In general, the people most severely affected by an eventual flood and living close to the vulnerable portion of a road or an embankment contribute to these initiatives with their voluntary labour, regardless of the fact that people who are not contributing may also benefit from their work. In such moments of crisis people are not much concerned about whether their benefit is proportional to their contribution.

The largest number of initiatives belonging to this category has been found in Chaptir Haor. In ten cases they are relatively minor and recurrent practices in those portions of the embankment that are regularly cut by the *beel* leaseholders to facilitate the final fish catch. As mentioned earlier, there is a consensus among farmers and leaseholders about this practice, provided the leaseholders fill the breaches after having completed their fishing. The leaseholders, however, generally fulfil their obligations to close the cuts only to the extent that no breach will occur under normal circumstances. Every year, as soon as the water level in the river rises, the farmers check the filled-in spots and, if necessary, raise and strengthen the closures.

These initiatives involve the work of only 10-20 farmers for about one day, but they effectively protect from flood hundreds of acres of land and in some cases the whole *haor*.

Two initiatives in Chaptir Haor stand out for their material and organisational dimensions. They relate to a 1996 embankment breach which threatened to destroy the crop of the whole *haor*. Thousands of people worked for two nights and a full day to raise two roads that served as compartmental embankments. Thanks to these initiatives, a dramatic and tragic event that could have caused a severe local food shortage was successfully prevented as only a small fraction of the crop got destroyed.

Three similar initiatives were found in Beel Singri, where high levels of the Mahananda River threaten to flood the entire area almost every year and to damage standing *aus* and *aman* crops. About 475 labour days were spent on these works in 1997. The initiatives successfully protected an area of about 3,500 acres of land on which the livelihood of seven villages depends.

In Banskhali a similar initiative was taken in 1995 by the people of Shilkup Union, after a portion of the coastal embankment was severely damaged. This caused the intrusion of saline water, which badly affected agricultural lands, ponds and homesteads. The initiative consisted in closing the breach and repairing a 350 m section of embankment. About 100 people, including 80 volunteers and 20 labourers paid by wealthy farmers, worked for ten days. Funds were collected to provide food and entertainment (which included the rental of music equipment) to the labourers. This endeavour protected several homesteads and about 300 acres of agricultural land from the intrusion of saline water, thus benefiting about 2,000 people. The embankment also serves as a road and thus has an important role for transport and communication. For the last two years, the concerned embankment section has been regularly maintained and repaired by the local population. However, it continues to be a vulnerable spot and people fear that they may not always have the capacity to protect the embankment without external assistance.

An initiative to repair and maintain a portion of an embankment in Galachipa concerns a very vulnerable location in the southern part of the polder, where a few years ago the BWDB constructed a 'retired' (set-back) embankment, as it was not considered possible to continue rehabilitating the original embankment. This left 63 families and an area of about 75 acres of land in an extremely vulnerable situation. About 10 families abandoned their homes and lands and migrated. The remaining families, having no better option, decided to try their best to continue to obtain protection from the abandoned embankment. Given the high risk to which they are exposed, they do not wait for the rainy season to take action; every year during the dry season all adult male members of the 53 families jointly work on a voluntary basis to rehabilitate the embankment. About 500 labour days are required to resection the 350 m embankment. So far, they have been successfully maintaining the old embankment, thereby preventing the loss of their homesteads and lands to river erosion.

Case 13 — People's Collective Effort to Raise a Dike near Baishaki Khara

In March 1996 a breach in the submersible embankment surrounding Chaptir Haor threatened to destroy the only paddy crop of the entire *haor*. The people of Chaptir Haor were aware for quite some time of the vulnerability of the embankment portion that eventually breached. The two weeks that preceded the breach, when the water level in the river began to rise more than usual for that time of the year, several large farmers and *matabars* visited the BWDB office almost every day to call their attention to the emergency and seek their assistance. At the same time they already started with their own means to protect the embankment with earth, sand bags and bamboo mats. A group of men patrolled the embankment day and night to watch the situation. Restrictions were imposed on navigation because the waves caused by boats could have further weakened the embankment. The public water transport that could not be deviated to other water routes was given signals to keep the maximum possible distance from the embankment to prevent the waves from hitting it.

The day before the breach, some BWDB officers visited the spot, assured the concerned farmers that the embankment would not breach, and left again. The people of Chaptir Haor, however, did not trust the judgement of the BWDB engineers. Already a few days prior to their visit, in the context of a meeting called by some *matabars*, people started to prepare for the emergency situation, if the embankment breaches. It was decided that while they continue to do their best to prevent the breach, they would also start to prepare themselves for a failure. They further discussed the possibility of raising a *janghal* (contour bundh) crossing the *haor* from east to west, to the north of the weak embankment spot. This would possibly protect about 80 per cent of Chaptir Haor's crop. Immediately after the meeting, three days prior to the breach, about 300 people, under the leadership of a few large farmers, had started to raise some low portions of this *janghal*. The breach, however, occurred on the evening of 23 March, 1996. Most people who had spent their day there strengthening the vulnerable embankment section with sand bags, bamboo mats and earth had already returned to their homes. Only a dozen people remained to watch the situation. When the embankment breached they kept trying to prevent the water from entering. One person ran to the nearest mosque and by microphone appealed to all the people to come immediately to the location of the breach.

However, in no time the breach became so big that they realised that no human effort could prevent the water from flooding the *haor*. All 50 mosques were immediately requested to announce through their microphones that all men should go with spades and baskets to the location. An Islamic conference that was taking place that day was interrupted due to this crisis and mosques played an important role in mobilising the people. Indeed all 50 mosques in Chaptir Haor repeatedly announced the message for several hours; and within less than two hours the 1.5 km *janghal* was covered with people. About 4,000 people worked without interruption for the whole night and the following day. Kerosene lamps, torch lights, dry food and *biri* (a local type of cigarette) were provided free by wealthy farmers and shopkeepers.

The following evening, the people felt assured that they would be able to protect the entire crop beyond the raised road. Accordingly, the number of people decreased to around 1000. In shifts, they continued working on the *janghal* for the following three days. Two days after the breach a meeting was called by a group of *matabars* and large farmers to discuss the situation. The meeting was publicly announced through microphones and all people were invited to attend. Each village sent a delegation of two to three representatives, so 93 people attended the meeting. They discussed the causes and consequences of the breach; and they all agreed that the breach occurred due to BWDB's negligence. A major disaster was averted entirely through their own initiative. The participants further discussed forthcoming expenses and, among other things, the need to recruit a group of watchmen who would patrol the embankment day and night until the end of the harvest season. They agreed to employ, at a daily wage of 100 Tk,[14] 15 watchmen for ten days and five for a full month. They estimated that around 55,000 Tk were required to pay the watchmen, purchase lamps and kerosene, rent a microphone (to call the people in case of an emergency), and other minor expenses. The amount that each village was expected to contribute was based on an assessment of its economic situation. To this aim the meeting selected three elderly men who were known and respected by everybody and requested each of them to prepare a list of all villages and indicate the amount that, according to their judgement, each of them should contribute. The financial contribution to be made by each village thus consisted of the average figure of these three independent estimates. Each village was formally requested to come up with its contribution to the fund as soon as possible. All representatives agreed with this procedure and assumed responsibility to mobilise the corresponding funds in their own villages. They did not need to raise the money directly from the people through a door to door campaign, as almost all villages had a common *panchayat* fund.[15] In addition, some wealthy people personally could and did contribute several thousand taka each.

During the meeting people also expressed their concern about the future. They decided to form the Embankment Protection Committee (*Badh Rakkha Samiti*),[16] which was entrusted with

---

14. This daily wage was about double the locally prevailing daily wages of agricultural labourers, which showed the enormous importance that was given to this task.

15. This institution, unique to the *haor* area, is discussed extensively later in this chapter.

16. From then onwards the *janghal* had the vital role to provide primary flood protection. Accordingly, it is no longer considered a *janghal* but a *badh*, which means a dike or embankment.

the responsibility to take all necessary actions to protect and maintain the structure in the future. One to two persons from each village became members of the committee's executive board. A chairman, a secretary and a treasurer were unanimously selected among those members who during the previous days had shown strong commitment to the cause. In 1997, the year after the breach, the Committee executed periodic maintenance works on the embankment at a cost of 60,000 Tk, collected, as in the previous year, from all villages in proportion to their estimated benefit. The *Samiti* has also tried to receive support from external agencies such as the Union Parishad and the BWDB, but so far has not been successful.

Salt inlets, Ring bundhs and a Partition Dike

These three types of hydraulic systems — salt inlets, ring bundhs and partition dikes — all relate to Banskhali's salt production and shrimp culture. A total of 270 acres of land in the study area have long been used as salt fields. Out of these, 205 acres currently are used also for shrimp culture. Although salt and shrimp are produced on the same land, these two trades have a different history and involve two distinct categories of people. Salt production takes place exclusively in the dry season and shrimp culture, during the monsoon. Salt production started around 30 years ago in a small portion of land and gradually increased. It mainly involves relatively poor and landless people leasing small plots of land from owners on an yearly basis. The annual lease for one acre of land is 16,000-18,000 Tk, which is almost three times the amount paid for leasing paddy on the land. Salt production is relatively profitable, though it is cumbersome and labour intensive. About 500 people are directly involved in the production of salt in the 270 acre region.

Shrimp culture started in Banskhali in 1984. The first shrimp scheme only covered 45 acres of land. Between 1984 and 1996 the area under shrimp production increased to 140 acres; and in 1997 an additional 65 acres were brought under this scheme. Shrimp culture is profitable but capital intensive. It is controlled by less than 20 rich people. The largest production unit, covering an area of 140 acres of land, is controlled by a group of 13 people. Five among them, together with three other associates, are also in control of the

65 acres of land which were brought under shrimp culture for the first time in 1997. Due to the fact that this land would otherwise have remained fallow, the lease amount they agreed to pay to the landlords was much lower than that paid by the salt producers. Initially the shrimp producers promised to pay 5,000 Tk per acre, but they deducted 2,500 Tk as contribution to the expenses related to the construction and maintenance of the required water management infrastructure.

In fact, the productions of salt and shrimp depend on the construction of specific water management structures, which need to be carefully operated and maintained. The shrimp producers had to invest considerable amounts of money for the installation of pipes, the construction of cement inlets, ring bundhs and a 1.5 km partition dike. These works were all done by paid labourers and together cost more than 500,000 Tk. In those areas exclusively under salt production, a number of BWDB sluices are operated and maintained by small informal groups of people, as per the requirements of the salt producers. Their O&M activities consist, for example, in the forceful opening of flapgates and the construction of ring bundhs near defective sluices to retain saline water. For these services they charge the salt producers an 'O&M fee' of 250 Tk per acre. As may be noticed, water management initiatives related to salt and shrimp production are closely linked to private enterprises led by people who have the power to construct their own infrastructure and to modify and take over the management of public infrastructure as well, when necessary to serve their own businesses.

Collective Purchase and Management of State-provided Irrigation Schemes

As was discussed in the previous chapter, until the beginning of the 1980s most minor irrigation technology was owned and operated by government agencies. When the government began the implementation of its privatisation programme, farmers were given the opportunity to purchase the irrigation schemes, generally at fairly low prices. According to the government's official policy, priority for the purchase of irrigation schemes was to be given to farmers' cooperatives and only if these did not show any interest were they to be sold to private individuals. It already was well known that cooperatives often were controlled *de facto* by private individuals. So in most places individuals, not groups, benefited from privatisation. In Beel Singri we came across some of Bangladesh's oldest LLP schemes, which until recently had been owned and managed by the BWDB. Contrary to what occurred in many other areas of Bangladesh, most of these schemes today are owned collectively.

Ten BWDB LLP schemes along the Mahananda River had been constructed by the BWDB in the early 1960s. Each scheme included lined field channels, pump houses, storerooms and other facilities and employed a number of BWDB staff. Although in those days the farmers paid almost no irrigation tax, this was not perceived as an advantage.[17] The management of the schemes was so poor that their command areas did not exceed 40-50 acres. Moreover, the yields of the irrigated crops were low due to a constant shortage of water. The privatisation of the 10 Beel Singri LLP schemes started in 1984 and was completed in 1990. As a result of this property transfer the command areas of the schemes increased from 40-50 acres to 150-200 acres.

The price paid by the farmers for each scheme ranged from 100,000 to 200,000 Tk. Out of 10 schemes only one was purchased by a private individual. The remaining schemes were purchased collectively by all farmers owning land in the command areas and all are managed today by executive committees. The funds to purchase each scheme were raised on a share basis among the farmers in proportion to the land owned. Committees consisting of 5-11 members were formed for the purpose of managing each scheme. A driver, a fieldsman and a manager — paid by the farmers — were appointed in each scheme.

17. No water tax was collected from the farmers until 1972, when the government officially introduced a service charge of 100 Tk per acre. In line with this new policy, the BWDB began to employ in each scheme a *patwari* (tax collector). However, in practice the farmers rarely paid any tax to the government, either because the *patwari* was not particularly motivated to collect the taxes from the farmers or did not have the authority to enforce payment or did not consider it necessary to forward the collected revenues to the government, a problem known throughout South Asia (see Small *et al.* (1989), Duyne (1994b), SRP/BWDB (1993).

Before the irrigation season begins a committee holds a meeting to prepare an O&M budget and to fix the water charges for the coming season. The corresponding amount is collected from the farmers, who receive payment receipts. Monthly meetings, open to all farmers who want to attend, are held during the irrigation season to review and monitor all the activities of the scheme. A general meeting is held once a year to inform the farmers about the financial status of the scheme and any future plans. If any farmer complains about his crop being damaged due to excessive water, he is exempt from the payment of water charges. Maintenance costs for field channels and electric motors are included in the water charges. Per acre water charges range from 800 to 1,200 Tk for the cultivation of *boro* and 600-700 Tk for *aman*. No conflict regarding water distribution and operation of these schemes was found. The farmers are well organised and express their satisfaction with the present management, which they perceive as honest, fair and transparent.

Case 14 — Collective Management of an Ex-BWDB Irrigation Scheme

The LLP scheme lifting water from the Mahananda River was constructed by the BWDB in 1962 to provide for irrigation of dry season crops. BWDB staff, including one driver, one *kalashi* (sluice operator), one *patwari* (tax collector) and one extension overseer were employed to ensure its O&M. A committee consisting of 11 farmers was formed by the BWDB. One person was engaged by the farmers as an irrigation supervisor. For this task he got in kind remuneration of 15 kg of paddy for each irrigated acre of land supervised. The BWDB supplied the fuel and also covered all other costs related to O&M. During the BWDB period the management was very poor. The pump was never operated for more than eight hours per day and the crops always suffered a shortage of water. As a result, the farmers were not interested to increase the area under irrigation. The farmers were dissatisfied with the operation of the scheme and complained about misuse of funds and BWDB staff's lack of responsibility.

The LLP scheme was purchased in 1989 for 105,000 Tk by a group of farmers of Palsha and Srirampur. The group was led by Mr. Azizur Rahman. The scheme included six motors, one pump, a pump house, a storage room and 1 km of lined channels. At the time it was handed over to the farmers, only the pump and one out of six motors was operational. The funds to purchase the scheme were raised among the 50 farmers cultivating land in the potential command area at the rate of 2,000 Tk per share. The command area of the scheme at the time it was handed over to the farmers was only 20 acres but gradually it got increased to 175 acres.

The scheme is managed by a committee consisting of 11 members who employ a manager, a secretary, a cashier, two drivers and three irrigation supervisors. The manager, the secretary and cashier, each are paid 1,000 Tk per month, the driver 800 Tk; and the three irrigation supervisors are each paid 150 Tk per acre of cultivated land within the scheme section under their control. The committee chairman is not getting any financial remuneration. At present all six motors are working. All farmers are currently very satisfied with the management, with the quality of the service and with the irrigation charges.

In 1996 the committee fixed the water charges at 1,200 Tk per acre for the *boro* season and 600 Tk per acre for the *aman* season. Each shareholder who contributed 2,000 Tk during purchase of the scheme is getting free irrigation for one *bigha* of land. If any land is affected by water logging, its owner is compensated by the management. The total scheme provides irrigation for the cultivation of 175 acres of *boro* and 125 acres of *aman* paddy. The lands are cultivated by more than 102 farmers.

Cleaning Canals and Embankments

In the course of this study the people of Galachipa recalled that a few years ago they participated in large numbers in an initiative to clean the coastal embankment and an important canal. This collective endeavour to clean a major canal, clogged with water hyacinth, and to support the Embankment Maintenance Groups (EMGs)[18] in clearing

18. Embankment Mainenance Groups consist of 10-15 poor women employed by the BWDB to ensure the preventive maintenance of embankments. The programme used to be funded by the government of the Netherlands with the dual objective of providing employment for poor women and enhancing the maintenance status of BWDB's embankments. Though the programme proved to be viable from a social, economic and technical perspective, it was discontinued after the withdrawal of Dutch funding (see Duyne (1994c)).

the embankment slopes of shrubs and bushes, was not a local initiative in the strict sense, as people were motivated by the staff of a water development project. It is briefly discussed here because it proves that people, if they support and understand the benefits of a project, they are also willing to make significant contributions to exogenous initiatives.

The initiative was promoted in 1993 by BWDB's *Systems Rehabilitation Project* (SRP) staff in collaboration with the Union Parishad when it was realised that the newly formed EMGs would not be able to clean the embankment without additional support. In fact, since the embankment was completed almost 20 years earlier, it had never been maintained at all. Clearing its slopes of thick bushes and shrubs was beyond the capacity of EMGs. The local BWDB staff discussed the issue with the UP chairmen and then organised a meeting with the local population, who expressed their willingness to assist the EMGs in this task. Accordingly, about 250 people participated in a two-day clean-up of a 25 km stretch of embankment.

A similar initiative was taken to solve the problem of drainage congestion caused by an excessive amount of water hyacinths in an important drainage canal, a problem that affected about 1,700 acres of land elsewhere in the same region. In 1994, the farmers of Ratandi were encouraged by the SRP team to remove the water hyacinths and some 500 people participated on a voluntary basis in this action.

Operation and Maintenance of BWDB Regulators

This section discusses people's capacity to manage exogenous water management technologies, with particular reference to the BWDB's regulators and related infrastructure. Most BWDB regulators were constructed without consulting the local people about their requirements and without seeking their future involvement in O&M. In most FCD systems, there are hardly any funds for maintenance and no operation plans. Even when a *kalashi* (sluice operator) is officially in charge of operation and sluice committees or water users organisations are set-up within the framework of water development projects, the actual O&M plans often remain vague and the practices are not necessarily consistent with the system's formal objectives.

Although there have been local initiatives to operate and maintain most BWDB regulators, the data presented in this section are primarily based on case studies of O&M practices, in ten out of 35 regulators in Banskhali and four regulators in Beel Singri.

In both Banskhali and Beel Singri, the most striking finding related to the operation of BWDB regulators was that, while they were designed for a certain purpose, they often ended up being used to pursue entirely different objectives. For example, 35 regulators constructed by the BWDB between 1967 and 1988 in Banskhali were all equipped with flapgates on the assumption that people would not have any particular interest in controlling the water levels inside the area protected by the polder. This, however, is rarely the case. Once people are provided with a technology that can potentially enhance their control over water levels, they are likely to take advantage of this opportunity. In Banskhali, for example, in many locations people forcefully open or close the flapgates for several reasons: to allow the intrusion of saline water for salt and shrimp culture, to allow the intrusion of sweet tidal water for irrigation and to retain water for agriculture, for fish-culture or for other domestic water uses.

In Beel Singri, the regulators were constructed with the objective of protecting the *aus* and *aman* crops from flood and facilitating post-monsoon drainage of lowlands to enhance the production of *boro*. While among the farming community there is general consensus on keeping the regulators closed to protect homesteads and fields from the early floods, drainage is a more contentious issue. In fact, in the dry season many people would like to use this new technology for retaining water in *beel*s and *khals,* water that could be used for irrigation and domestic purposes. Such a use of regulators deviates considerably from what they have been planned for. However, in an area characterised by severe dry season scarcity of surface water, and where groundwater is not affordable for the vast majority of small farmers, the preoccupation with maximising storage of surface water is quite understandable.

People's capacity and willingness to maintain water management infrastructure is a subject debated among policy-makers, technocrats and international experts. The local population is often blamed for the poor maintenance status of regulators, even when no users' organisation has been given the official mandate to ensure O&M. When looking at the poor status of some regulators, external visitors often ask — provided it is important for their livelihood — why people do not take better care of this infrastructure. This study shows that people assume maintenance responsibilities of exogenous technology according to their own criteria and capacity. For example, if a sluice is leaking and it is beyond their capacity to repair it, they often construct crossdams or seal the leakage with earth in order to prevent the passage of water. In Banskhali, where regulators are considered vital but have never been adequately maintained, these informal maintenance practices are well-established. In some cases, Sluice Committees have been formed by informal leaders or UP chairmen. Four sluices in one study area are operated and maintained by an informal committee serving the interests of the salt and shrimp producers. In another case, where the operation of a sluice was causing severe conflicts, local people decided to hand over its management to a local youth club that was recognised by both parties as being impartial. In all these cases, those who assume responsibility for sluice management collect service charges of the order of 120-240 Tk per acre from the people who benefit from their endeavours.

Case 15 — Operation of Paglakari, Dhukumari and Chachania Regulators in Beel Singri

The Paglakari and Chachania regulators were constructed by the BWDB in 1985 within the framework of a project funded by the Canadian government. This was perhaps the first project in the water sector in Bangladesh requiring local participation in sluice operation in the context of FCD systems. The proposed institutional framework, however, was inadequate. In fact, the project institutionalised the formation of 'Sluice Committees,' whose chairmanship was generally given to the local UP chairman and the secretary was a BWDB official. The remaining four members were selected by the BWDB from among the rich and influential farmers of the village nearest to the sluice, even if its command area covered dozens of villages. Sluice committees thus were either inactive or were hijacked by the local elite, whose water management interests did not necessarily coincide with those of the majority of the people.

In Beel Singri a six-member Sluice Committee was formed for each regulator. The person who was selected by the BWDB to become chairman of Paglakari regulator was Mr. Alamgir Khan, a former UP chairman and a powerful informal leader. Mr. Khan belonged to an old feudal family and was respected and feared in the whole area. Every day people of all classes visited him to seek his advice or assistance on various issues. Mr. Khan's connections with government officials made him an important patron, an intermediary between ordinary people and the state. His cousin, Mr. Nurul Khan, was selected chairman of the Chachania Sluice Committee.

These two persons virtually control not only the operation of Paglakari and Chachania but also the Dhukumari regulator, where no formal Sluice Committee was ever organised. The two cousins owned extensive portions of very low land in the command area of the regulators, land that can be brought under cultivation only if there is effective drainage. Though such massive drainage is consistent with the formal objective of the FCD system, it does not correspond to the will of the large number of small farmers who own medium and high land and who would like to maximise water retention in *beel*s and *khal*s for irrigation, cattle washing, and domestic purposes. However, given the fact that both Sluice Committees were controlled by those favouring drainage, the preference of the majority had little influence on the operation of the sluice.

Although many people were openly grumbling about this situation, the undemocratic management of Beel Singri's sluices never led to any open conflict. When Beel Singri was selected as a pilot project by a Dutch-funded water development project, its staff immediately detected the socially 'illegitimate' nature of the Sluice Committees. A two tier organisational structure, consisting of Water Users Groups at the village level and an elected Water Users Committee at the system level were supposed to replace the autocratic Sluice Committee. Surprisingly, however, Mr. Khan again became the head of these organisations, elected this time through a democratic process. Although it seems possible that this institutional change might have led to a somewhat more transparent and predictable operation of Beel Singri's sluices that kept into account the needs of medium and highland farmers, a considerable volume of valuable surface water, which many people would like to retain for the dry season, continues to be drained.

This example shows that the introduction of exogenous democratic institutions may have little influence on local water management practices and that in cases of conflicting interests, those who have more power are likely to have the upper hand. It also points to an important difference between irrigation systems and flood control and drainage systems. Whereas in irrigation systems basically all farmers want to use the infrastructure for the same purpose, in FCD schemes there are many distinct categories of stakeholders with often incompatible interests. Thus, as shown in the case presented below, no matter how a system is operated, among the affected people there are always going to be winners and losers.

Case 16 — Conflict-ridden Operation of a Fulbari Regulator (Beel Singri)

In 1984 the BWDB constructed a water control structure at Fulbari which would allow for control of the water levels of Beel Lakshmikol. It is not clear whether BWDB's objectives were to facilitate drainage or water retention or simply to enhance human control over water. The majority of people living and owning land around Beel Lakshmikol welcomed this technology, which would allow them to retain in the *beel* more water for irrigation. The Fulbari regulator, as was discussed earlier, triggered the construction of nine relatively large gravity flow irrigation canals that were designed, constructed and entirely financed by groups of farmers led by LLP owners. Thus, 245 acres of land that used to be fallow were brought under *boro* cultivation. Over 200 cultivators consider themselves benefited by the Fulbari regulator. Only one person, Mr. Mohidul Islam, is vehemently against using this regulator for water retention. Mr. Islam is a lawyer and absentee landlord, living in Chapai-Nawabganj, who owns about 30 acres of lowland near Beel Lakshmikol. Apparently Mr. Islam inherited this land from his father, who had received it as a gift from the *zamindar* for his services as a *kabiraj* (traditional healer). The land used to be fallow until a few years ago, when Mr. Islam constructed a small dike around his land. The dike allows him to bail out the water and cultivate one *boro* crop. After the construction of the regulator, which was used by local farmers to retain water, he was no longer able to cultivate his land because of the high water level favoured by the majority of the Beel Lakshmikol farmers. In order to get his way, Mr. Islam filed a court case against the LLP owners, the Fulbari Sluice Committee and the BWDB. Through his professional connections he was able to influence the court, which decided against the interest of the numerous local farmers. In fact, a legal order was issued that the BWDB was responsible to operate the Fulbari regulator in such way that the water level in Beel Lakshmikol would not exceed three feet. Mr. Islam personally employs a watchman to control the operation of the Fulbari regulator. After this court decision the number of LLP schemes stagnated and the command area of the existing schemes decreased.[19]

Case 17 — Conflicts over Management of Sluices in Banskhali

Although most sluices in Banskhali are equipped with flapgates and officially do not require any operation, the local population immediately understood that by applying some force, they could open and close the gates according to their requirements. In several cases, however, there has been profound disagreement on the purposes for which these regulators should be used. In

19. Conceivably it was Mr. Islam's right to protect his land from being inundated. However, as was mentioned, Mr. Islam's land, like that of many rich landlords who own large portions of low land, prior to the construction of BWDB's infrastructure, used to remain under water and could not be used for cultivation. According to the still valid 1950 State Acquisition and Tenancy Act, which was intended to abolish the *zamindari* system, feudal landlords were not entitled to retain water bodies greater than the size of a tank (Januzzi and Peach 1980). Accordingly, low marshes officially became government property. BWDB's hydraulic infrastructure was often explicitly intended to dry these lands without ever addressing the critical issue of related property rights. Many people in Beel Singri claim that the large landlords who, these days, claim ownership over lowlands have appropriated them illegally, which is likely to be correct. It may also be that once these marshes became cultivable their former owners (or their heirs) were able to claim back — more or less legally — their entitlements. The impact of public hydraulic infrastructure on property rights and in particular its contribution to the privatization of commons is an important but sensitive issue, which thus far has not been studied as yet.

two cases local entrepreneurs discovered that large amounts of fish could be caught near the regulators by opening the gates and allowing tidal water to enter. Such a practice, however, causes the submersion of agricultural fields and damages the paddy of several farmers. Although the farmers are very upset about these 'fishery schemes,' they are not able to protect their interests. In one case the owner of the fishing scheme felt so confident about his control over the sluice and the adjacent *khal*, that a few years ago he invested 16,000 Tk for the construction of a permanent cement dam. In this way he created a small reservoir, in which he has started commercial fish culture. These two fishing schemes have forced the farmers in the surrounding areas to construct their own drainage system by installing pipes through the embankment.

Another sluice was controlled for several years by a group of salt producers. They used to manipulate the gates in such a way as to allow the penetration of saline water. The farmers surrounding this salt scheme were upset about this practice, for it damaged their crops, but they were too weak to oppose it. In 1995, however, the situation changed. In the context of the LGED-IFAD Bamerchara Small Irrigation Scheme (BSIS), the farmers got organised into a formal committee. This committee is formally recognised and supported by the Union Parishad and several governmental organisations. This has allowed it to take control over the sluice and has forced the salt producers to stop manipulating the gates for their own benefit.

6

Socio-Economic and Organisational Dimensions

Distribution of Benefits

THE previous chapter has described actions taken by individuals or groups of people to improve their control over local water resources. When referring to the people involved in or affected by local water management initiatives, however, few details were given about the socio-economic status of those who benefited and those who may have been negatively affected by these initiatives. Given that water and related resources may be considered Bangladesh's most important common property resource and that unequal access to productive resources is typical of rural society in Bangladesh, it is crucial to assess how different strata of the society have been involved in these initiatives. A distinction first needs to be made between 'actors' and 'stakeholders' in water management, that is, between those who are directly involved in an initiative and those who, though affected, are not active participants. In terms of the social distribution of associated benefits, local water management initiatives may be divided into four broad categories:

(a) *Initiatives aiming at enhancing or protecting agricultural production, which benefit all farmers cultivating land within a certain area, regardless of their socio-economic status, and do not have any negative impact on other groups of people*

Irrigation canals in all sites where they could be found and Chaptir Haor's *janghals* and embankment closures belong to this category. Irrigation canals benefit farmers within their command areas, regardless of the amounts of land owned. As they are linked to surface water bodies far from homesteads, the use of water for agricultural purposes does not compete with domestic uses, though it may compete with fishery. *Janghal*s are not only functional to farmers. Many are used as roads by the whole *haor* population. However, given the fact that agricultural land in rural Bangladesh is distributed very unequally, with some 50 per cent of all households being completely landless, it is quite clear that the main actors and beneficiaries of these initiatives are those households who own or cultivate land. Nevertheless, common interests among both large-scale and small-scale farmers and their capacity to unite to provide themselves with a collective good are facts that have been underestimated by many social scientists. This study has found that cooperation between wealthy and powerful farmers and poorer ones not only is possible but also that rich farmers often make disproportionally large contributions to collective goods. Collective action in water management in rural Bangladesh confirms Olson's observation that the distribution of the burden to provide a collective good is rarely proportional to the benefit gained, and that, contrary to typical Marxist assumptions, there is a "systematic tendency for 'exploitation' of the great by the small" (Olson 1965:29)

(b) Initiatives which benefit all people within a certain area, regardless of their occupation and socio-economic status, and which do not have any negative impact on other groups of people

Several crossdams in Beel Singri, Chaptir Haor and Galachipa and most embankment cuts and drainage outlets in Banskhali, belong to this category. In Chaptir Haor, where tubewells are scarce and costly to install, a number of fairly large crossdams are constructed on the initiative of village leaders to satisfy domestic water requirements of all villagers. These benefit all people within a certain area, regardless of their gender and socio-economic status. Considering that several wealthy people in Chaptir Haor own private tubewells, it may be argued that the crossdams particularly benefit less privileged people.

However, the majority of the reservoirs created through crossdams, though they are also important for domestic uses, are primarily intended to enhance the availability of water for agriculture. This primary function has priority over domestic use, even if numerous people would opt for the latter. Those people who depend on surface water for their domestic requirements are the socially weaker ones, those who do not have the power to influence decision-making processes related to the allocation of surface water. In Beel Singri, where during the driest months many tubewells fail to supply water, some poorer people said that they would favour saving the water stored in crossdam reservoirs exclusively for domestic uses but they are not in a position to influence such a decision. Nevertheless, the interests of the landless poor in this domain are ambivalent. As agricultural labourers they are employed in the fields of farmers who could intensify production by appropriating a disproportionate share of water, so they do have some reasons to consent to agriculture-promoting behaviour. This came out clearly in a study about the social distribution of benefits of BWDB's Beel Singri FCD system, which found the landless to be among the most ardent supporters of what Rostow would call economic 'trickle down' theories. In their opinion, the introduction of modern hydraulic infrastructure and irrigation technologies has had a tangible impact on demand for local labour. During seasons in which they formerly were forced to migrate they now can find employment locally. The increased labour demand, they argue, also has had an impact on local wages, which have shown a tendency to increase. The indirect benefits derived from increased employment opportunities and wages may compensate for poor people's lost share of domestic water resources and thus may contribute to their acceptance of the situation.

Embankment cuts and the installation of drainage outlets, like those found in Banskhali, may be primarily motivated by the need to release drainage congestion from agricultural fields. However, these cuts usually benefit the homesteads of all people living within their command areas regardless of their socio-economic status.

Typical of these initiatives is that they benefit relatively large and heterogeneous groups of people, but some people more than others; and those who are most benefited often bear the full cost of an initiative. In some cases this may be explained by the high transaction cost that would be involved in collecting contributions from all beneficiaries and by the difficulty of forcing material participation on the loosely organised group gaining only secondary benefits.

(c) Initiatives which benefit people (or a specific category of people) within a certain area but which have a negative impact on groups of people outside the area

Several crossdams belong to this category of initiatives. They are appreciated by all people living near the water reservoir created by the crossdam but they may have a negative impact on people from other areas or may deprive them of any access to water. A typically negative impact often caused by crossdams is drainage congestion. To avoid this problem and the social conflicts to which it may lead, most crossdams are removed at the end of the dry season and reconstructed after the monsoon. Drainage congestion caused by crossdams is a problem found in Galachipa, where many are constructed by individual families or small groups of people to increase the availability of surface water for domestic use and to create ponds for fish culture. Unlike other regions, in Galachipa the seasonal reconstruction and removal of crossdams are not customary.

In Beel Singri, as illustrated by Case 4, while all people living in the area surrounding the reservoirs created through crossdams consider themselves benefited by these initiatives, people whose lands and homesteads are located upstream are negatively affected by the high water level, which is a major impediment to crossing the canal by bullock cart. In Banskhali, some of the crossdams constructed for irrigation purposes are resented by people living downstream along the canal. They feel that it deprives them of their share of water. When rainfall is scarce and supplementary irrigation for the *aman* crop becomes a necessity not only to obtain better yields, but also to prevent the crop from dying, resentment may escalate into violent conflicts. This occurred in Banskhali some years ago, in relation to a crossdam constructed by a group of farmers to irrigate their fields. Due to a severe drought the farmers downstream from the dam got so upset, that they violently attacked the upstream farmers. Several people got injured and one farmer was killed. A number of crossdams built near Banskhali's regulators with the objective of retaining sweet water for irrigation or saline water for shrimp-culture cause drainage congestion to land located upstream.

Whether or not negatively affected, people take action to protect their interests. These depend on a number of factors. First of all, it was noticed that people generally do not go beyond grumbling if the damage inflicted on them is not too severe. This is particularly the case if the initiative occurs outside their territory. The situation may suddenly change, however, as indicated by the violent reaction of people upstream Beel Singri's canal (Case 4) and by the clash between upstream and downstream farmers in Banskhali, if the negative impacts become too severe.

Secondly, whether or not those negatively affected by an initiative react violently or not depends on their relative power in relation to the person or group of people behind the initiative causing the problem. If the people associated with an initiative are powerful, those negatively affected may not be able to do anything beyond raising a complaint with the local Union Parishad, which may or may not take action to solve the problem. However, if those negatively affected have some power or manage to gain the sympathy or active support of local authorities or external agencies, they may be effective in taking action. For example, the cut of the BWDB crossdams on Bowalia Khal in Galachipa was led by a UP chairman and was strongly supported by the whole population (Case 10). The relative power of the actors behind a local initiative also influences the awareness and attitude of government agencies towards them. The response of local BWDB officials to embankment cuts, for example, appears to be influenced by the socio-economic status of the actors involved. As was illustrated with reference to Chaptir Haor's embankment cuts, the BWDB nowadays regularly closes the cuts made by the economically and politically powerful *beel* leaseholders, but it does not close some of the cuts made by farmers.

In Banskhali one embankment section was regularly cut by a small farmer whose lands were badly affected by water-logging. Some years ago he was caught on the spot by a BWDB officer, who took him to his office and threatened him with legal prosecution. Once the farmer persuaded the officer of his reasons to cut the embankment, he was forced to agree that he would bear the full cost related to closing the cut and repairing the damaged section of the embankment. This appears to be a reasonable agreement that also could have been made with other people involved in similar practices. However, far different standards were applied to a rich and influential person who constructed a permanent cement dam across one of BWDB's *khals* to start commercial fish culture and to the shrimp farmers whose water inlets and manipulation of BWDB's regulators are perhaps even more damaging. As may be expected, there are consistent rumours among the local population about government officials being bribed to ignore these practices, a relevant issue that will not be further pursued in this study.

The power-balance between various stakeholders may change as a consequence of exogenous factors, such as, for example, the official recognition by government agencies of the violated entitlements of a specific group or individual. This was illustrated by the case related to water management practices in Beel Lakshmikol (Case 16) and by the case related to the management of some of Banskhali's sluices (Case 17).

Another factor influencing people's response to water management practices affecting their livelihood is the geographic distance between the site of the problem and that of the water management initiative itself. It may happen that negatively affected people are not even aware of the cause of their problem. This was found to be the case with a number of crossdams in Galachipa that caused drainage congestion in other areas. In general, even when people are aware of the cause of a problem, it is difficult for them to react to something that occurs beyond their immediate surroundings.

(d) *Initiatives which benefit a specific group of people or private individuals and may or may not have negative impacts on other stakeholders*

All the initiatives related to salt and shrimp production in Banskhali belong to this category. The construction of such water management infrastructure has been part of a process which transformed what used to be paddy fields into salt and shrimp fields. Although in this area — unlike other parts of Bangladesh — there is no widespread resentment about these industries, the crossdams built with the objective of retaining saline water and the operation of BWDB regulators according to the needs of salt and shrimp producers, have a negative impact on the adjacent agricultural lands. On the other hand, the costly compartmental dikes constructed by shrimp producers to prevent sweet water from the hills from entering the shrimp fields are equally important to protection of the farmers' paddy fields from the intrusion of saline water.

Many of the embankment cuts in Chaptir Haor were done by leaseholders to facilitate the final fishing (Cases 9 and 11). This is a traditional fishing practice, which happens to be functional within certain limits, also for farmers. To safeguard their collective interest, the *haor* farmers have developed socially recognised rules to ensure that the interests of an individual or a specific group of stakeholders do not have a negative impact on other people: leaseholders must ask permission prior to cutting an embankment and are accountable for their water management practices to the local *matabars*.

To summarise the analysis of who benefits and who is negatively affected by endogenous water management practices, the first issue that may be noticed is that most initiatives are closely related to agricultural practices. While they are often neutral for non-farming households, they benefit all those whose livelihoods depend on agriculture, including landless tenants and labourers regardless of their socio-economic status. However, all initiatives that consist in withdrawing substantial volumes of surface water for agriculture as well as those practices directly aimed at fishing have a negative impact on fish resources and thus contribute to the decreasing availability of fish, which negatively affects the poorer section of the rural population in particular.

Quite a number of initiatives found to have a negative impact on farmers were related to fish culture. It is important to realise that the people involved in fish culture did not belong to the traditional fishing community. In these days fishery and fish culture are very profitable activities, mostly controlled by affluent people. Traditional fishers do not have the power or material resources to control surface water bodies and related infrastructure and are mostly employed as labourers. Indeed, the interests of the traditional fisher community appear to be systematically neglected by local and governmental water management initiatives alike.

An observation made in the course of this study is that women — at least in the visible aspects of endogenous water management initiatives — were completely absent. None of the labour days spent on local initiatives was performed by female labourers. Not a single woman was found to be a member of any of the informal committees formed in relation to local water management initiatives or to be present in meetings related to their planning and implementation. In some cases, however, women were intensively involved in a supporting capacity. For example, in the case of Chaptir Haor's endeavour to raise the *jangal* to protect their crop from flood, women contributed by sending snacks and sweets to the men engaged in earth works. This was an essential contribution, because the emergency situation did not allow men to return to their homes to eat. Women's contribution was instrumental in keeping the work going and was perhaps a way to express their solidarity. In general it was found that women were aware and well informed about water management practices taking place in their areas. The main reason for the absence of women in local water management initiatives is that these take place in public spaces, which in rural Bangladesh remain primarily the domain of men. Though not involved as actors, women, who are responsible for the provision of drinking water, for washing and cooking, certainly must be considered as principal stakeholders in relation to local water management practices. If, for example, excessive water is withdrawn from surface and groundwater sources, women's domestic water sources may dry up, forcing them to walk farther distances to fetch water and bathe. Nevertheless, in general, it may be said that women perceive their water management interests as being basically identical to men's.

Another aspect that needs to be addressed while analysing the benefits and negative impacts of endogenous water management practices in rural Bangladesh is their long-term effect on public infrastructure and on the natural environment. People's endogenous solutions to the drainage congestion caused by BWDB's embankment in Banskhali, although in the short-term benefit and thus supported by all people within a certain area, may threaten the sustainability of a vital embankment and thus increase the vulnerability of large numbers of people. Many collective actions described in this study indicate that people in rural Bangladesh are not generally concerned about the preservation of natural resources and that some practices are rather problematic in terms of environmental sustainability. Chaptir Haor's and Beel Singri's ingenious strategies to mobilise resources for maintaining indigenous hydraulic infrastructure (and incidentally to finance the construction of a mosque) indicate that local communities — like the government — tend to manage their fish resources with the only objective being to generate revenue. This already has contributed to the depletion of availability of fish. The excessive drainage or withdrawal of surface water with its negative impact on the recharge of aquifers is facilitated by modern irrigation technologies, but people themselves take numerous initiatives to make them more effective. It was noticed that people, when talking to external agencies about an actual or proposed water management initiative, think of their own immediate needs without necessarily keeping into account long-term consequences either for themselves or for others. This should be kept in mind by the development practitioners and participation advocates who appear to believe that local participation in water development planning will inevitably lead to equitable and sustainable solutions.

Patterns of Resource Mobilisation

Most people in rural Bangladesh do not pay any taxes. In the water sector, a number of attempts to introduce O&M charges have been made in the context of irrigation systems. All of them failed. The high incidence of poverty, the protracted dependence on aid and rich people's notorious reputation as free riders are often used to explain people's presumed lack of capacity and willingness to pay. This study shows, however, that people do have the capacity and the will to contribute substantial amounts of money, labour and material for the development, maintenance and operation of all types of water management infrastructure. Indeed, some of the goods provided through locally mobilised resources involve costs that exceed by far the development budgets of Union Parishads.

This section examines how groups of people with a common interest in the provision of a specific good overcome one of the most critical collective action problems, namely the mobilisation of the required human, material and financial resources. To this aim, a distinction will be made between 'collective goods' and 'public goods.'

MOBILISING RESOURCES FOR 'COLLECTIVE GOODS'

Collective goods benefit a clearly defined and occupationally homogeneous group of people. This means that they serve only one purpose for one specific category of stakeholders. Typical collective goods are the provision and maintenance of distinct types of irrigation systems (those created through canals, crossdams or inlets), small *janghals* and drainage outlets. With regard to the prevailing patterns of resource mobilisation associated to the provision of collective goods, a distinction may be made between four types of situations:

(*i*) *Collective goods to which all group-members are expected to contribute with cash or labour in proportion to their benefit*

Contributions in proportion to the yielded benefits are the most common type of practice in relation to the provision of and maintenance of irrigation canals in Beel Singri, Banskhali and Galachipa. In these cases an estimate of costs is made by the leader of an initiative, after which the total amount is divided among the beneficiaries. The contribution of each individual may vary from a few Taka to several hundred Takas. In Beel Singri, for example, farmers contribute to the construction of the irrigation canals around Beel Lakshmikol with average contributions ranging from 120 to 150 Takas per *bigha* of land owned within the command area. These cases indicate that the majority of the people have a preference for cash contribution, though in all regions it was found that poorer group-members were sometimes given the opportunity to contribute labour instead of cash. The preference for cash may be explained by the fact that (a) in rural Bangladesh landowners who care about maintaining a certain social status would never engage in earth work; and (b) people pursue other economic activities besides agriculture.

Even if people contribute to the provision of a good with shares proportional to their personal benefit, they still depend on the initiative of one or a few leaders who bear the full transaction costs associated with the collective action. These cases support Marwell and Oliver's theory, that collective action depends on a 'critical mass' of highly interested and resourceful individuals rather than on 'average' group members (Marwell and Oliver 1993).[1] These people have a higher interest than other group-members in the provision of the collective good and are therefore willing to take upon themselves the burden of mobilising funds, recruiting labourers and supervising construction. In Beel Singri the group's 'critical mass' consists of the LLP owners, who, besides being interested in irrigating their own fields, intend to sell water to other farmers. In irrigation schemes not associated with pumps, action leaders usually are farmers who own larger than average amounts of land under the scheme.

(*ii*) *Collective goods to which all group-members are expected to contribute but not in proportion to their benefit*

The maintenance of Chaptir Haor's irrigation canals and small *janghals* is ensured through a different approach. First of all, people participate with labour rather than cash. Those who consider manual labour below their social status send their domestic labourers. Secondly, according to the local informal rule, the system users are expected to participate by sending one labourer per household for as many days as are required to complete the works regardless of the amount of land owned. Considering the unequal distribution of land, this norm does not appear to be very fair. In practice it was noticed, however, that the system works somewhat differently. Those who own only a marginal portion of land inside the command area of an irrigation system do not necessarily comply with their formal obligations or they may participate in the works on the first day only; whereas larger farmers assume most maintenance responsibilities. Given the fact that it would be rather complicated to convert into labour time the share of benefit of each group-member, particularly because the volume of work varies from year to year, the norm regulating farmers'

1. See chapter 2.

participation in maintenance works increases the likelihood that each system user contributes at least one day of labour.

Also in this case, it may be noted, group heterogeneity in terms of unequal interest among group-members in the collective good is closely associated with socio-economic inequality, but has a positive impact on its provision. If all farmers benefited equally, none would be willing to make up for free-riders. In fact, the collective actions associated with the maintenance of Chaptir Haor's small *janghals* and irrigation canals indicate that free riding is not an unknown phenomenon. But in the case of local water management practices those who benefit most make up for those who fail to participate, doing so apparently without resentment. Another reason why wealthy farmers often assume disproportionately high shares of maintenance responsibility may be that rich households in the *haor* area employ on a permanent basis quite a number of domestic labourers, who, during the maintenance season, are not very busy with other work.

(iii) Collective goods provided by one person or a small group who collect from the beneficiaries a 'service charge'

Some endogenous water management systems identified in this study are entirely financed by private individuals who recover their costs by charging service fees to the people benefiting from them. This type of provision-mechanism could be found only in Banskhali and was in most cases associated with the provision, operation and maintenance of hydraulic infrastructure related to salt or shrimp production. The entrepreneurs involved in these schemes made considerable financial investments to build the infrastructure they needed. Their cost-recovery strategy consisted in charging the owners of the land under shrimp culture 2,500 Tk per acre. They deducted this sum from the 5,000 Tk per acre landrent, which they originally agreed to pay. Considering that the infrastructure was essential for their enterprise rather than for the landowners, this practice may be seen as a way to push down the landrent. The same strategy to provide a collective good also was found in relation to a number of crossdams built near leaking regulators to retain water for irrigation.

Though these types of provision patterns have the flavour of little private enterprises, it needs to be said that earning money by selling the associated service was not the prime motive behind these initiatives. Indeed, the service charges only allowed the initiators to make a minor profit on their investment without calculating the value of the time they themselves invested in the initiative. Also in these cases the initiators were primarily motivated by their personal need for the infrastructure and in some cases by the 'by-products' (fish) which they could obtain from these surface water irrigation systems. The reason why this method of providing a collective good was only found in Banskhali may be the high degree of commercialisation of the local economy and the fact that even more than in other regions most households are involved in addition to agriculture in off-farm jobs. Charging a 'tax' to whoever wants to use the service (for example, by installing a LLP on a reservoir created through a crossdam) is an effective way to keep transaction costs low for it eliminates the often time consuming need to collect contributions from all group-members.

(iv) *Collective goods provided by one person or a small group of people who do not attempt to mobilise contributions from other beneficiaries or to recover 'service charges'*

This situation was found in all regions though more frequently in Banskhali in relation to the provision of some drainage pipes and in Chaptir Haor in relation to embankment cuts and closures and the construction of crossdams (as described in Case 5). The reason why a few individuals take upon themselves the burden and expenses required by the collective goods are several. In Banskhali, though the purchase and installation of drainage pipes is associated with relatively high costs and labour and generally benefits numerous households, people who take upon themselves this burden are those most severely affected by drainage congestion. This type of action is thus associated with a heterogeneous group whose members have an unequal interest in the good being provided. In relation to drainage outlets, however, it is not the 'poor exploiting the rich' (which is the case in the maintenance of Chaptir Haor's *janghals* and irrigation canals); rather, it is those who are less negatively affected by water-logging taking a free ride on those who are most affected. Nevertheless, to use Olson's concepts again, the drainage outlet providers have a 'selective incentive' to bear the full costs of this action for they retain the exclusive control over the fish that can be trapped at the drainage outlet. This incentive (which in the case of larger, collectively financed inlets is considered as remuneration for the person responsible for their O&M) together with the fact that these initiatives are sufficiently small to be affordable by one or a few individuals, may explain why the providers willingly accept the cost-burden.

The initiatives belonging to this category in Chaptir Haor reflect different situations. In one case (Case 9) they concern the embankment cuts executed by the *beel* leaseholders. The cuts also benefit the farmers. The costs involved in cutting an embankment are minor, particularly in relation to the enormous benefit in fish that the leaseholders can obtain through the practice. Moreover, in order to catch the fish and prevent people from stealing them, the leaseholder somehow needs to keep his employees on the spot. Accordingly, the participation of all beneficiaries in cutting the embankment is not the issue. The main collective action problem associated with this water management practice is the preservation of the resource system. This is ensured by informal but strictly enforced norms and rules, which give agricultural requirements priority over fishery and accordingly entitle farmers to decide about the water levels to be maintained in *beels* and *dhobas*.

In Chaptir Haor, a number of crossdams intended to retain water for irrigation were entirely financed by one rich farmer, who, though allowing unrestricted access to other farmers, neither sought their contribution nor attempted to charge them any 'water tax.' This behaviour did not so much reflect the rich farmer's generosity as it did the disproportional benefit that he derived from the crossdam. Once more we are confronted with what Olson defined a 'privileged group,' in which one or more of the members have a sufficient interest in the good to make it worthwhile taking upon themselves the full burden associated with its provision. The reason why similar situations may not emerge, for example, in Banskhali is that the region is characterised by many small farms. Accordingly, none of the group-members have, in comparison to other group-members, a similarly strong interest in the good. Large landholdings such as those found in Chaptir Haor are also found in Beel Singri. In this area, however, rich farmers, besides owning many different types of land, also have access to groundwater. This indicates, as pointed out by Hardin (1982), that heterogeneity may either enhance or dissipate the prospect of collective action. Whether rich farmers have the possibility of getting the good from another source, such as, for example, through the purchase of an STW, they may choose this alternative instead of collective action even if the associated costs are higher.

MOBILISING RESOURCES FOR 'PUBLIC GOODS'

'Public goods' are goods with multiple functions, serving heterogeneous groups of stakeholders who cannot be excluded from their benefits regardless of whether they pay for them or not. Examples of public goods are embankments (and related efforts to repair or raise them), large *janghals* covering lands of more than one village and used as roads, crossdams aimed at retaining water for irrigation and domestic use or those constructed for domestic use only. Such locally provided public goods are found in all regions, though they are significantly more numerous in Chaptir Haor than elsewhere. The mobilisation of resources for the provision of a public good follows a different logic, depending on whether it relates to a sudden emergency or is a recurrent event with predictable resource requirements.

(i) *Mobilising resources to face emergencies*

All four regions covered by this study sometimes face emergencies or high risk situations associated with embankment breaches or erosion, both of which can have devastating effects on lands and homesteads. Even though a breached embankment is a state-provided infrastructure that should be maintained by the government, some embankments are so vital that people cannot usually wait until action is taken by an external agency. Such emergencies are recurrent events in many locations in Chaptir Haor, but they also have occurred at least once during the last few years in the other three regions.

Typical of these emergency situations is that many people participate with labour and resources without efforts being made to distribute the burden in proportion to the benefit. However, once the acute crisis is past, there is a tendency of wealthier people, who do not engage themselves in manual work to provide the required material and financial resources and of poorer people to participate with their labour. This was the case, for example, in Banskhali, where a local effort to repair the embankment involved about 100 volunteers for 10 days. Wealthier people mobilised among themselves resources to pay for labourers for 20 days besides provision of snacks, fruits and local cigarettes (*biri*) for everyone and the rental of music equipment to provide a more pleasant working atmosphere.

In Chaptir Haor, in cases of emergency, people are called by the village leaders through the mosque's microphone, a situation taken sufficiently seriously by everyone to participate with their plain capacity (see Case 13). Once the acute crisis is over, if additional work is required, this may be done for by the village funds, an important resource, discussed further below.

(ii) Recurrent provision and maintenance of public goods and services

Several local initiatives relate to infrastructure with multiple functions. For example, while the primary function of a crossdam may be to retain water for irrigation, the same water may also be used for bathing cattle or for domestic purposes. In addition, a crossdam may also have an important role in communication and transport. Thus, while in terms of its primary function the infrastructure is a 'collective good' it may be considered a 'public good' for its auxiliary functions. In such cases the costs related to the initiative are generally borne by the primary beneficiaries following the systems described for the provision of collective goods. This, however, also implies that those group-members who ensure the provision of the good feel entitled to decide upon the use of the resource system, which explains why irrigation needs systematically get priority over domestic ones.

The only region characterised by the massive presence of vital endogenous hydraulic infrastructure that may be considered a 'public good' and that accordingly had the need to develop viable institutions to ensure its provision and maintenance is Chaptir Haor. These institutions indicate a profound local awareness of the intrinsic difference between collective and public goods and of the potentially high transaction costs involved in collective action of large groups. In fact, whereas small collective goods are provided and maintained directly by their users, the maintenance of public goods is ensured through budgetary allocation of funds by the *panchayat*, an extraordinary institution unique to the *haor* area.

(iii) *Role of external resources*

A number of water management initiatives may be called 'mixed initiatives'. These have been partly funded by locally mobilised resources and partly by external resources (e.g., FFW wheat or Thana Parishad funds). Mixed initiatives were found in Beel Singri and Galachipa to finance the construction of crossdams. Attempts to receive external support for locally identified projects are regularly made in Chaptir Haor, where it was found that, once in a while, the endogenous *janghal*s are rehabilitated under the Union Parishads' FFW programmes. A substantial volume of maintenance work, such as, for example, repair of embankments and closing of breaches and cuts, is in some years done by the local people and in others by the BWDB or other government agencies.

The allocation of external resources to the provision or maintenance of endogenous hydraulic infrastructure depends on stakeholders being able to persuade or involve the Union Parishad chairmen. This is more likely, if the chairmen are locally perceived as honest and if they themselves are stakeholders in the initiatives.[2] Some cases indicate that Union Parishad chairmen may have the unique capacity and legitimacy to mobilise both local resources and external resources. In Galachipa a large dam built almost thirty years ago across Bowalia Khal, to which the Thana Parishad and the local population contributed with almost equal shares, is still considered by all people to be one of the most fundamental development projects in the area. At present most local infrastructure is supplied by line agencies within the framework of internationally funded development projects; so agency staff tend to be more interested in persuading donors about the need

2. A study on local governance found that from a representative sample of UP chairmen, about half were perceived as 'relatively honest' by villagers. In these Unions, potential was detected for local resource mobilisation and improved service delivery by UPs (Barenstein (2000b)).

to finance costly projects than in making themselves accountable to the public by mobilising local resources. Line agencies thus tend to bypass locally elected bodies, further undermining their ability to fulfil their mandates.[3] Accordingly, the role of the Union Parishad in the provision of local public goods is rather marginal. On this subject, a recent study on local governance found that,

> . . . UP officials do not (and could not) do much in the way of municipal services (as done elsewhere by local governments in rural areas), their importance to the life and livelihood of the majorities diminishes, and with it, their capacity to mobilise resources and involve people in their projects.
>
> — Barenstein 2000a:145

VILLAGE FUNDS

The mobilisation and administration of collective funds for specific purposes, such as the construction or maintenance of a mosque or a primary school, is common all over Bangladesh. Such funds are generally mobilised by members of an informal or registered mosque or school committee formed by the village population at large, by the rural elite or by specific groups of stakeholders (e.g., the parents of the pupils). However, a village-based institutionalised approach towards generating funds on a continuous basis and for multiple purposes appears to be unique to the *haor* area.

Village funds are found in most of Chaptir Haor's villages. Some refer to them as *samajik funds* (society funds), others as *panchayat fund* (collective funds); and in the Hindu villages they are known as *debatta fund* or *mahadeb fund* (gods' funds). Most common and understood by everyone in the *haor* area is the term *panchayat fund*. Nobody recalls the origin of this institution but the tradition of

3. Officially, there is a long list of functions to be performed by the UP. They include the "adoption and implementation of development schemes in the fields of agriculture, forest, fisheries, livestock, education, health, cottage industries, communication, irrigation and flood protection… ." (GOB 1983: Art. 30). The large discrepancy between these prescriptions and local realities is also due to "the very low resources at their disposal, and their dismal performance in the collection of revenue." (Barenstein (2000a:245)).

collecting funds for collective purposes has a clear religious connotation. Particularly in Hindu villages people associate the origin of their village funds with what used to be the payment system of the *Siral*. The *Siral* was a person recognised by the community for having the mystic power to keep hailstorms away from the *haor*. Whenever the menace of a hailstorm would approach, he was urged by the community to go to the fields and pray to the gods to send the threatening clouds elsewhere. A tax of one *seer* of paddy per *keyar*[4] of cultivated land was collected among all farmers after the harvest. Half the paddy collected in this way served to compensate the *Siral* and half to finance various religious ceremonies. The last *Siral* died just before independence, around 1970. Today, either because the society has become less superstitious or because nobody appears to have such powers, no *Siral* has replaced him. Nevertheless, the custom of collecting resources for a collective purpose and particularly for protecting crops continues to enjoy the necessary ideological legitimacy.

Muslim villagers also associate the origin of *panchayat* funds with a religious tradition. According to their memories, their forefathers used to collect paddy after the harvest for the construction and maintenance of mosques and *madarsas*, a practice that is still going on today. In many villages rice is collected on Fridays from each household; and women are expected to put aside a handful of rice every time they cook. This rice, collected each Friday by a different person, is primarily used to remunerate Imams and to maintain mosques. Similar fund-raising practices associated with the maintenance of religious institutions exist almost everywhere in Bangladesh.

The *haor*'s village funds, however, do not have much in common with these traditions neither in the way they are generated nor in the way they are used. An analysis of the sources of income of five villages indicates that at present village funds come from various sources, the leasing of *dhobas* and the sale of fish from various surface water bodies representing the most common and important sources of

4. *Seer* is a local weight measure equivalent to 900 g. A *keyar* is a local land measure equal to 0.33 acres.

revenue. Most *dhobas* are regarded as *ejmali* properties, that means collective property. They have numerous owners but the owners do not claim their property rights.

While the collection of contributions from the general population has lost importance, private donations are very significant in some villages. Also common and rather important in some villages is the collection of informal taxes from shopkeepers. In one village the most important source of revenue is the annual organisation of a *mela*. A *mela* is a village fair offering various types of entertainment. The most important attraction in this particular *mela* is gambling. The organisation of *melas* requires the formal authorisation of the Upazila Parishad and gambling is illegal. In order to obtain the permission for their *mela* and to persuade officials to ignore this rather overt and often contentious fund-raising strategy, in 1996 the village *matabars* had to pay a bribe of 10,000 Tk to the police. As indicated in Table 6.1 one village controls a fund almost three times the average village fund and thus may not be particularly representative. In fact, private donations for two specific purposes (the construction of a mosque and a college) constitute the most important sources of revenue of this village fund. This village also succeeded in obtaining a financial contribution to its self-financed college from the Upazila Parishad. The combination of a few very dynamic local leaders, good connections with Bangladesh's political elite and several successful *Londonis* explain the exceptional situation of this village.

As indicated in Table 6.1, villages spend on average 37 per cent of their funds for the maintenance of hydraulic infrastructure, with one village spending almost 80 per cent of its funds for this purpose. Whereas even today an average 30 per cent of village funds are spent for religious purposes, this figure is influenced by two villages. One of them is a poor, entirely Hindu village, whose population belongs to the Fishers caste. This means that they do not have a great stake in hydraulic infrastructure designed to protect agricultural production. The other village received a donation of 800,000 Tk for the construction of a mosque from a family residing in London.

Table 6.1: Synoptic view of five village income and expenditure in Chaptir Haor (1993-1997)

Village	Average annual fund	Source of village funds (%)					Expenditure (%)					
		1	2	3	4	5	1	2	3	4	5	6
1	377,200	70.1	2.0	8.4	8.5	11.0	5.8	34.5	48.0	-	4.2	7.5
2	49,300	-	6.7	40.6	52.7	-	37.7	13.2	13.7	16.6	-	18.8
3	83,000	-	-	56.6	43.4	-	59.7	30.7	-	-	4.8	4.8
4	91,000	26.3	-	66.2	-	-	79.1	-	14.3	5.5	-	1.1
5	10,600	-	1.9	98.1	-	-	2.5	12.7	74.1	-	-	10.7
Average	130,700	48.2	2.1	54.0	20.9	2.2	37.0	18.2	30.0	4.4	1.8	8.6

Sources of village funds:

1. Private donations
2. Collection from villagers
3. Leasing of common and sale of fishing services
4. Taxation of shopkeepers, *mela* and gambling
5. Donations from Thana Council

Expenditures:

1. Hydraulic infrastructure
2. Education
3. Religious infrastructure
4. Collective water transport
5. Market and graveyard development
6. Savings

Source: Interviews with village leaders

There is a certain correlation between the proportion of money generated by means of leasing out common resources and the amount spent for the provision and maintenance of hydraulic infrastructure. This may be explained by the fact that — as already mentioned above — private donations are often intended for specific purposes, such as the construction of mosques and schools. Nevertheless, funds generated by leasing out *dhobas* also are used for purposes other than hydraulic infrastructure. For example, one village pays a lump sum of 13,000 Tk to the river *ghat* (boat landing wharf) so that the whole village population can cross the river by boat for free for a full year.

One village purchased a boat for 7,000 Tk and employs a boatman for 6,000 Tk to transport children to school. Due to lack of transport, the children of people who do not own private boats otherwise could not attend school during the monsoon period. Two villages also spent a considerable amount of their common funds for the construction of sanitary latrines in schools. The absence of these facilities was considered a major impediment preventing girls from attending schools. A peculiar investment aiming at improving the education in the *haor* was made by a village, which spent 9,000 Tk to purchase a television for the teachers of their newly constructed college. According to the *matabars*, some incentives beyond their salary are necessary in the *haor* area, where, due to the rather difficult living conditions, it is hard to attract qualified teachers.

The graveyard development works undertaken by two villages involved raising burial grounds above flood level. They used to be flooded during the *haor's* long rainy season, making burials in this period of the year almost impossible.

In order to raise the village market, one village combined these works with digging of a big pond. The fishing rights associated with this pond are now leased out on an annual basis, thus creating a permanent source of revenue for the village.

As shown in Table 6.1, at the time of this study each village controls a certain amount of savings. These savings are especially important during emergencies. For example, it is out of these savings that most villages make their contribution to the expenses related to raising the *janghal* after the breach in the embankment.

Village *matabars* are responsible for managing *panchayat* funds. No person is specifically appointed for this purpose except the treasurers, who keep written records of the income and expenditure of the village. The *matabars* are also in charge of organising the auctions of fishing rights in the *dhoba*. These auctions are announced publicly, are open to everyone and are truly competitive. Decisions about the use of the funds thus generated are made public during the same meeting. Unlike the central government-controlled *beels*, which generally end up in the hands of rich and powerful outsiders, the *dhobas* all are leased to local people, including groups of Fishers who fish themselves. They can pay the established sum in instalments but in any case they must complete payment within the month of September, when the village needs funds to rehabilitate its hydraulic infrastructure. Given that the auctions are public and that, for many villages, they represent the main source of revenue, most men have a rather precise idea of the total amount in their village funds. There is wide consensus about the way village funds are generated and spent.

Chaptir Haor's village funds share some remarkable similarities with the "standing funds" found by Wade (1988) in some villages of Andhra Pradesh, in southern India. As in Wade's villages, funds are primarily generated by means that are officially considered illegal or by public auctions of communal resources; and they are used to finance the provision of public goods required to sustain the local production system. The existence of village funds, like those found in Chaptir Haor and some Andhra Pradesh villages, is not mentioned in any other study on India or Bangladesh. Whereas Wade was unable to say whether they exist in other regions of India, in the course of this study we systematically checked whether a comparable institution could be found in the other three regions. This allows us to state that village funds do not exist in other regions of Bangladesh.

Organisational Practices

As was discussed in this chapter, in terms of labour, local water management initiatives may involve any number of people: from a few for a couple of hours up to thousands for several days. Regardless of the number of people involved, the initiatives may have a very local impact or they may affect the livelihoods of thousands of people. They may happen only once or may reflect an indigenous water management practice occurring regularly since time immemorial. Whatever the case, most initiatives are carried out by groups of people who do not feel the need to establish a specific formal organisation for that particular purpose. Considering the importance given in development circles to the formation of local level organisations, this is a significant observation.

One of the reasons why the establishment of formal organisations may not be considered necessary is that most initiatives are associated with agricultural practices and involve an already well-defined group of farmers. This is especially true in the case of the collective provision and maintenance of irrigation canals, which make up 56 per cent of actions reviewed by this study. Typical of these groups is that they are relatively small and that they consist of people who know each other and who are all dependent on the same source of water. These groups of farmers cultivate adjacent plots that are all located within an area defined and delimited by clear boundaries. This situation may explain why they do not need to set up any organisation to manage their collective activities. Indeed, if we consider that one reason to form an organisation is to reduce any tendency to free ride and shirk, this is not required for a group of people who already know each other or any group which is small enough to make such behaviour noticeable. Further, as pointed out by Olson (1965), the transaction costs associated with setting up an organisation increase with the size of the group. Most collective actions presented in this study are associated with relatively small groups. Beside the fact that individuals' failure to participate would be noticeable, they tend to pursue goals, which are so important to group members, that each is willing to participate even if he has to contribute a higher proportion of the costs involved.

However, some collective actions occurring occasionally in all regions and being regularly pursued by the people in Chaptir Haor, involve relatively large groups or at least what Olson would classify as intermediate groups. According to Olson, intermediate groups are not able to obtain a collective good without some group coordination or organisation. Yet, people in rural Bangladesh appear to be rather successful in the provision of some public goods even without setting up any informal or semi-formal organisations. This is particularly striking in Chaptir Haor, the area with the strongest indigenous water management culture and the most ancient, varied and complex hydraulic infrastructure.

The absence of specific organisations in charge of water management does not mean, however, that the *haor* society did not develop its institutions to cope with the challenges posed by the local environment. The normative system regulating the management of its common property resources — such as the *dhoba*, the organisation of *panchayat* funds, the informal rules about providing voluntary labour for the maintenance of water management infrastructure or the social preparedness for natural disasters, which allows to mobilise thousands of people within a few hours — proves that *haor* society as a whole is organised for the management of its vital infrastructure. *Haor* society, depending on the issue at stake, is subdivided into villages or hydrological units, whereby, depending on the situation, an individual may participate and be a stakeholder at different organisational levels. Thus, one of the reasons why formal organisations with the specific mandafe to operate and maintain water management systems are generally not considered necessary even by relatively large groups, is that people in rural Bangladesh are already part of a highly structured and organised society, in which everybody is aware about his or her rights and obligations in relation to common resources. At the village level everybody knows the amount of land and other resources owned by other villagers, and thus how much each household is expected to contribute to a certain initiative. Indeed these groups have all the characteristics that Singleton and Taylor (1992) attribute to 'communities', which, according to these authors, are able to solve collective action problems endogenously and with a minimum of specialised institutional roles.

When local initiatives are of public interest, key roles are often played by the village *matabars*. The authority of these informal leaders is respected within the village and also in neighbouring villages. In general it was noticed that decisions such as fixing the date for the annual maintenance works of canals and *janghals* are made and announced by persons who are implicitly recognised as leaders within a certain area. In many cases decisions are taken in the context of casual but inevitable encounters in the fields, in the market or in the mosque. In other cases, village *matabars* or others with a large stake in a certain initiative call a meeting to decide on common actions.

The social organisation of rural Bangladesh is effective not only in cases of relatively small initiatives but also in cases of very large ones, such as raising the *janghal* in Chaptir Haor (Case 13). Even though a semi-formal organisation eventually was formed for that particular initiative, this was done only after the crisis was over; and it had the objective of ensuring the sustainability of the initiative. Another reason why most water management initiatives, even recurrent ones, do not lead to the formation of any type of specific organisation, is that most have an impact on a rather limited area belonging to a group of people who already know and spontaneously interact with each other whenever necessary.

Nevertheless, a few informal and semi-formal organisations have been set-up in relation to collective water management practices. Besides the *Badh Rakkha Samiti,* informal or semi-formal organisations were found in relation to the nine ex-BWDB LLP schemes in Beel Singri and in the context of a number of water management initiatives in Polder 64/1b. In the Beel Singri LLP schemes along the Mahananda River, these organisations are dynamic cooperative societies responsible for managing technically complex irrigation systems. These organisations, however, are only active during the irrigation season. Their general members are not very much involved and do not expect anything beyond a good irrigation service. Although the management of these schemes is in the hands of relatively large farmers, they have shown that water management infrastructure and organisations are not inevitably captured by the rural elite for their private interests. When farmers were given the opportunity to purchase minor irrigation systems under the privatisation programme, all except two were bought up by the farmers collectively, even without external pressure and control.

A general tendency observed in most initiatives regarding leadership is that the majority of the people appear to feel comfortable delegating management responsibilities to their "representatives" and participating in formal meetings only once a year at most. In Banskhali the organisations formed to manage small irrigation schemes or operate some BWDB regulators were completely passive for years on end. They were completely unknown to the local people. Even people who were supposed to be members of these committees were no longer sure about who the other members were and what their positions were. This indicates that it is not only organisations formed by external agencies, such as the numerous water development projects that after some time may lack any clear mandate or activity. The same can happen in locally organised groups.

While informal or semi-formal organisations tend to have a rather marginal role in local water resource management, there are some key figures in rural Bangladesh who tend to have an important role whenever a certain initiative is of public interest. These are the traditional rural elite, who are locally known as *matabars*. *Matabars* are recognised as leaders by the people of a certain *para* (neighbourhood) or village, for their age, family background, leadership skills, formal education, management capacity, wealth and their access to government officials. They were found to have a positive role, especially when organisational skills or access to external resources were required. The role of *matabars* was particularly significant in Chaptir Haor, where we found the highest degree of consensus and harmony in relation to water management. Indeed, though the *matabars* had a certain role in all research areas, it was only in this region that people, if asked about the roles of *matabars*, would spontaneously mention the mobilisation and management of *panchayat* funds and related activities. In all other regions, even if *matabars* occasionally had a leading role in relation to a specific initiative, people only mentioned their role as members of the *salish*, i.e., in conflict resolution. In cases of conflict between different categories of stakeholders (e.g., salt producers *vs*. farmers or domestic *vs*. agricultural uses of water), they were generally not very effective in finding a solution and tended to be biased towards the most powerful and wealthy parties.

In several initiatives it was found that the Union Parishad had played a key role. For example, if people were not satisfied with the O&M condition of a certain BWDB scheme or when two parties had a conflict regarding the operation of a regulator, they often appealed to the Union Parishad chairman. This is not surprising, considering that the Union Parishad is currently the only locally elected body and the only formal link between the people and the central government. In a number of cases Union Parishad chairmen were found to have key roles in identifying and leading local initiatives; and they proved to have a unique capacity to mobilise both external and local resources for a specific goal.

To conclude, the analysis of organisational practices associated with endogenous water management initiatives confirms, as was discussed in Chapter 2, that under certain conditions, participation in the provision and maintenance of collective and small public goods does not require establishment of any specific formal organisation.

7

Contextual Conditions for Collective Action

The Singularity of Chaptir Haor

THE previous two chapters have shown that all over Bangladesh people have the capacity to cope with their environment and to organise collectively to pursue common goals. Nevertheless, the data presented also reveal strong regional differences in the nature, frequency and degree of institutionalisation of local water management initiatives. Chaptir Haor, the smallest among the four research areas, stands out as having a remarkably high number of local initiatives. In fact, 75.7 per cent of all cases presented in Table 5.1 (Chapter 5) relate to collective water management practices found in Chaptir Haor. In contrast, Beel Singri, the largest among the three research sites, accounts for only 4.5 per cent of the cases. The remaining 20 per cent of the cases are distributed in almost equal numbers between Banskhali and Galachipa. Chaptir Haor and the other research areas differ in other respects, not just in their water management practices.

First, whereas many initiatives found in Banskhali, Galachipa and Beel Singri represent local responses to state interventions, Chaptir Haor's indigenous water management initiatives all pre-date modern state interventions.

Second, most initiatives beyond the *haor* basin are undertaken by small groups in order to provide themselves with *collective* goods. The *haor* inhabitants, however, have developed a time-tested capacity to mobilise thousands of people and substantive resources for the provision of *public* goods. This capacity is used not only to sustain indigenous hydraulic infrastructure but also to provide villagers with other valuable goods and services, such as public water transport, public sanitary latrines and school facilities.

Third, several initiatives in Beel Singri, Banskhali and Galachipa, though beneficial for a particular category of stakeholders, have had negative impacts on other stakeholders. They are therefore associated with discontent among certain sections of the population or even with overt conflicts between distinct categories of stakeholders. This, however, is not the case in Chaptir Haor, where endogenous water management practices rest on a strong social consensus.

Finally, a remarkable difference between Chaptir Haor and the other three regions covered by this study, consists in the degree of institutionalisation of endogenous water management practices and in the role of the *matabars* in safeguarding public interests. In fact, it is only Chaptir Haor's initiatives that occur within the framework of informal but transparent and socially recognised rules. And only *haor* villages have developed mechanisms to mobilise local revenues required to sustain indigenous hydraulic infrastructure on a continuous basis. More specifically, Chaptir Haor appears to be the only region of Bangladesh where many villages have developed some corporate institutions similar to those found by Wade (1988) in some villages in south India.

How can the singularity of Chaptir Haor be explained? Why has Chaptir Haor succeeded, similarly to some of Wade's villages in South India, in sustaining a public realm, whereas Beel Singri, Banskhali and Galachipa have not? Providing some possible answers to these questions requires a closer look at the specific context and a comparison with conditions in the rest of Bangladesh.

The Haor Basin

The *haor* basin is located in the north-east of Bangladesh, south of the Indian Meghalaya and Assam hills and covering an area of about 4,500 km^2 (Rashid 1991: 24). It constitutes the country's largest inland depression, characterised by vast areas of land having an elevation of less than 3 m above sea level. The origin of this large bowl-like area seems to be connected with still occurring tectonic movements. According to Rashid (1991) the land sank between 9 and 12 m in the last 200 years; and it is estimated that during the last 50 years alone it has sunk between 0.9 and 1.5 m.

In current Bengali, *haor* is defined as a large saucer-shaped inter-fluvial depression surrounded by a perimeter of river levees. Most *haors* have deep central areas, called *beels*, which are permanently or semi-permanently under water. The word *haor* originates from the Bengali *shagar*, meaning sea and refers to the fact that for about six months a year such areas are deeply flooded and thus assume the aspect of a limitless sea. The whole basin encompasses 47 major *haors* and dozens of smaller ones. Like each individual *haor*, the basin as a whole has the shape of a shallow saucer, and remains under water for several months a year. During the dry season green paddy fields cover virtually all land. The basin is criss-crossed by numerous tributaries of the rivers Surma and Kushiara, which discharge the waters from the Meghalaya Hills, one of the heaviest rainfall areas in the world (12,000 mm per year). During the monsoon season, due to insufficient gradient, these runoff waters, combined the area's own rainfall of about 4500 mm a year, cover virtually all land to a depth of several meters.

The *haor* basin was among the last areas of Bengal to be colonised. Until less than a century ago much of its land was covered with reeds and shrubs and formed a key habitat for fish and waterfowl. Nevertheless the region attracted successive weaves of human settlers for several centuries. According to Rizvi (1970), the first settlements of ethnic Bengalis in the *haor* basin date back to the second century AD., which coincided with the period when Buddhism, with its aversion for animal food, became prominent in North Bengal. These early settlers were attracted by the basin's immense fisheries and confined themselves to fishing areas. Islam (1985) maintains that the first Bengali colonisers of this region were political refugees from the Barind region, Brahmins escaping from Muslim persecution. After having first settled in the north-eastern town of Sylhet, they moved towards the interior of the basin in the early fourteenth century, when the city was conquered by the Muslims. A second wave of migration brought to the *haors* numerous Afghani refugees, who escaped from their stronghold, Orissa, after it was conquered by the Moguls. These high caste Hindus and aristocratic Muslims brought with them a certain entourage of farmers and artisans; but numerically they were no threat to the sparse indigenous tribal populations, whose livelihood was based on shifting cultivation. Apparently, thanks to the abundance of land, these distinct ethnic groups co-existed for some time rather peacefully. The situation changed when the region came under Mogul dominance at the beginning of the seventeenth century. These new rulers systematically encouraged migration to the *haor* in order to increase their revenues from land and fisheries. The Bengali population thus began to grow rapidly and by the middle of the eighteenth century most cultivable land was brought under plough cultivation. This process eventually forced the tribal groups to recede to the northern sub-montane zone.

However, from 1780 onwards, due to a series of abnormally severe floods and earthquakes, there was significant out-migration; and by 1860 the area was nearly depopulated. At the beginning of the twentieth century the population density of then greater Sylhet District (which includes the largest portion of the *haor* basin) was only 412 inhabitants per square mile, which was less than half the density of the adjacent district of Comilla. Only from the first quarter of the twentieth century onwards, in relation to the increasing population pressure in the Bengal delta, did the *haors* begin to be gradually recolonised (Islam 1985, Rizvi 1970).

Chaptir Haor

THE PEOPLE

Chaptir Haor is a medium-sized *haor* located in the central and lowest portion of the *haor* basin. It covers an area of about 50 km^2; and with its 48,926 inhabitants it has a population density of approximately 966 person per km^2. Such populataion density is not only above the national average, but significantly higher than in Galachipa (718) and Beel Singri (333). It is, however, only one third of the population density of the heavily populated area of Banskhali (3292). Considering that Beel Singri is located within one of the driest regions of Bangladesh and Chaptir Haor within the wettest, these figures indicate a correlation between the availability of water and population density.

The present population of Chaptir Haor is primarily Muslim and consists of a mixture of people who claim to have settled in the area several centuries ago, along with people who migrated to the area over the last hundred years. About 18.5 per cent of Chaptir Haor's inhabitants belong to Bangladesh's Hindu minority, which is about 7 per cent above the national average (GOB 1994:132).[1] Some Hindus are farmers and large landlords but most are poor fishers who live in four villages near the main *beels* in the north-eastern part of the *haor*. All believe to belong to the first wave of migrants and thus originate from the north-western Barind region. As indicated in Table 7.1, Chaptir Haor's current population is scattered in 35 villages, belonging to three Unions of Derai Upazila.

Table 7.1: Chaptir Haor's Population

Union	*No. of Villages*	*Male*	*Female*	*Total*
Jogdhal	21	11,710	10,773	22,483
Karimpur	13	13,360	12,098	25,458
Tarol	1	505	480	985
Total	**35**	**25,575**	**23,351**	**48,926**

Source: GOB/BBS 1991: 71-75

Chaptir Haor's distance from Dhaka is 304 km and 114 km from Sylhet, an important provincial town. It is considered a remote area because it is not accessible by road even in the dry season. There is a road up to Derai, the Upazila headquarters, from where Chaptir Haor can be reached crossing the river by ferry boat. During the dry months internal communication in Chaptir Haor is ensured to pedestrians and oxcarts by a few earth roads. Rikshaws, which in rural Bangladesh are generally quite numerous, are hardly found in Chaptir Haor. All roads are submerged during the monsoon, when boats provide the only means of transport. In this period transport becomes altogether easier because each corner of the basin can be reached by boat.

1. According to the Population Census 1991 the national percentage of Muslims is 88.25. The percentage of non-Muslims is somewhat above the national average in all four regions.

Chaptir Haor provides limited employment opportunities beyond agriculture and fishery. Over 75 per cent of the *haor* population primarily rely on agriculture. About 44 per cent of the households report farming as their main source of income, whereas agricultural labour supports the livelihood of 31.3 per cent of the households. The role of agriculture in the local economy is similar to the one found in Beel Singri and Galachipa. As may be expected in an area with good access to an important urban centre, agriculture is less dominant in Banskhali, where only 34 per cent of the households report farming as their main source of income and 22.8 per cent, agricultural labour (GOB/BBS 1991).

Only 5.2 per cent of the households report fishery as their main source of income. The reason why the importance of fishery is not reflected in the *haor*'s employment structure may be that only those people who belong to the traditional fishers communities tend to admit that they derive their income from this low-status occupation, even though seasonal employment in the fishery sector may be equally important for other poor people.

Migration is an important economic strategy that is not reflected in Bangladesh's official population census. Chaptir Haor, located in Sunamganj District, belongs to what was known as the greater Sylhet Division.[2] This area has had very high rates of international migration for several decades. According to several key informants, there may be close to 600 people from Chaptir Haor working overseas and still keeping in close contact with their native home. International migration is common among upper, medium and even medium-low strata of the rural society but due to lack of capital, basically absent among the landless.

Whereas international migration is a viable economic strategy for those households owning land, temporary migration to some of the country's large towns represents an important survival strategy for poor people. Migration thus plays a pivotal role in the Chaptir Haor's rural economy. Migrants regularly send money to provide

2. Sunamganj became an independent district only in the early 1980s, as part of a general administrative reform, which led to the creation of 45 new districts.

support to those family members they have left behind, especially in periods of crisis. Chaptir Haor's emigrants not only support their family members. Some also send money to support collective initiatives such as the construction or repair of the indigenous hydraulic infrastructure, mosques and schools. In case of crop failure and during the slack summer season, many poor male migrate to Sylhet, Chittagong and Dhaka, where they mainly work as rickshaw pullers or construction labourers.

Natural Environment

Chaptir Haor is in many respects a typical *haor*. Being located in the central bottom of the basin, virtually all its land has an elevation varying between 2.6 m and 6.7 m above the sea level. During the monsoon period over 85 per cent of the land is flooded to a depth of several metres. In regular years the rivers overtop the levees and embankments surrounding the *haor* around the middle of May; and by the beginning of June the whole area becomes, in effect, a vast inland sea with only the villages and the highest river levees remaining above flood level. An almost complete reliance on deeply flooded lowland is perhaps the most outstanding feature of Chaptir Haor.

Chaptir Haor's villages are built on its *kanda*, the natural river levees encompassing the *haor*. This category of land, also called *athi,* constitutes only 5.2 per cent of Chaptir Haor's total surface. Most homestead land is naturally not sufficiently high to remain above the normal flood level and therefore is artificially raised about 2-3 m with earth. Accordingly, homestead land is very valuable and scarce. Only few vegetables and trees can be grown in the small areas of land high enough to be suitable for human settlements.[3] The borrow-pits from which the earth is taken are used as ponds and constitute an important source of water during the dry season. In fact, the groundwater table

3. Islam (1985), to whom we owe one of the only ethno-historical studies of the *haor* basin, makes a distinction between the slightly higher and wealthier 'vegetated villages' and the poorer 'naked villages' constructed on artificially raised lands. According to these taxonomies, basically all villages in Chaptir Haor would be naked, though — as indicated in the pictures — there is a significant difference in terms of amount and quality of homestead land owned by rich and poor households.

in the *haor* basin is relatively deep, and tubewells for domestic use are unaffordable for most people.

Over 80 per cent of Chaptir Haor's land is under plough cultivation. Cultivable land is locally called *abadi jami*, meaning reclaimed land. Contrary to other parts of Bangladesh, plots in this area are not delimited by little bundhs, as these would be washed away during the monsoon. Large extensions of land, locally called *bond* may be owned by several people whereby the portion owned by a particular person is either demarcated by stones or by *sangha*, a flood resistant grassy plant commonly used in lowland for this purpose. The lowlands at the centre of the *haor* have a somewhat higher value, for they are considered more fertile and are closer to surface water bodies used for irrigation. However, they are the first ones to be submerged in case of flash floods.

Table 7.2: Type of land owned by farmers (in %)

	Chaptir Haor	*Beel Singri*	*Banskhali*	*Galachipa*
Lowland	98	52	8	6
Medium lowland	2	4	28	14
Medium highland	3	38	64	72
High land	1	63	21	43
Total	**104**	**157**	**121**	**135**

Source: Soussan *et al*. (1998:23)

Because of Chaptir Haor's topography most cultivable land is single-cropped with *boro,* and even today local rice varieties are far more common than HYVs. Indigenous rice varieties have a maturation period about 20 days shorter than HYVs, which in an area prone to early floods may be crucial. Accordingly, HYVs are cultivated primarily by large farmers on about 9 per cent of the somewhat higher lands, which are irrigated with river water provided by means of LLPs. Only a decade ago these lands either remained fallow or were planted with floating paddy (broadcast *aman*), locally known as *bawa dhan*. HYV yields are much higher than those of *bawa dhan* but so are the costs of production and there is a high risk of losing the whole crop to flash floods.

Such homogeneous agriculture, which makes all people depend on a single crop, is not typical for Bangladesh. The contrast between the *haor* basin and other parts of Bangladesh becomes even more clear by looking at the great diversity characterising, for example, Beel Singri, where it is possible to find not less than 18 distinct cropping patterns. A rich variety of cropping patterns like the one found in Beel Singri with an even higher cropping intensity characterises Banskhali. Thanks to the proximity to Chittagong, Banskhali's cultivators produce vegetables under comparatively very profitable conditions. A rather limited agricultural diversification and a low cropping intensity is found in Galachipa, where, due to water constraints, most land remains fallow during the winter season. In Galachipa most land is single-cropped with rain-fed local varieties of transplanted *aman*. Only in about 17 per cent of the land is it possible to cultivate an additional *aus* crop, and in around 24 per cent, some rain-fed *rabi* crops. Nevertheless, though Galachipa's agriculture is rather backward and water shortages constrain its productivity, ever since it was provided with coastal embankments its agriculture has not been exposed to severe risks.

Another category of land found in Chaptir Haor is locally called *patit jami*. It consists of relatively high land located either in the south-west or in the north-east of the *haor*. This type of land is too high to be irrigated, and accordingly is used for grazing, as threshing ground for drying fish and nets and for recreational purposes. *Patit jami* can be either state or private property but regardless of its ownership, it is reserved for communal use. *Patit jami* is officially recorded as "fallow land" and covers about 10.2 per cent of the *haor*'s surface. Before LLPs were accessible to the *haor*'s wealthy farmers, this category of land was significantly larger. LLPs made it possible to lift water for irrigation from the river and farmers who prior to this opportunity allowed their *patit jami* to be used for communal purposes, now have brought it under cultivation. It is also said that much *patit jami* officially owned by the government had been illegally appropriated by the *haor*'s powerful elite.

Table 7.3: Land use in Chaptir Haor

Type of land use	*Area (ha)*	%
1. Gross area	5061	100
(a) *Uncultivated area*		
Homesteads, graveyards, kitchen gardens, schools, markets, roads, embankments, etc.	936	18.5
Forests and hills	-	-
Grazing land/permanent fallow land	518	10.2
Beels, ponds, khals,ditches and other permanent water bodies	154	3
(b) *Cultivated area*	4125	81.5
Single cropped area	4074	98.8
Double cropped area	51	1.2
Triple cropped area	-	-
2. Net cropped area	4125	
3. Total cropped area	4176	
4. Cropping intensity		101

Source: SRP/BWDB (1994b: annex (tables))

Land used for grazing is locally referred to as *beeler kanda.* It basically refers to the land covered by *beels.* It remains dry for about two months, after the water is being consumed for irrigation or drained for fishery. These types of grazing grounds are found mainly in the north-west, the location of Chaptir Haor's 17 large *beels.* Official figures refer to this category of land, which covers about 3 per cent of the Chaptir Haor's surface, as 'permanent or semi-permanent surface water bodies.' Even more than as a source of fodder, this category of land is important as a source of water for irrigation and as a natural habitat for fish. The availability of large portions of uncultivated land for communal pastures is unique to the *haor* basin. Elsewhere in Bangladesh, there is no communal grazing land other than the slopes of embankments and roads. Cattle are fed primarily in the homestead with crop residuals or they are allowed to graze in privately owned fallow lands.

Organisation of Agricultural Production

If Bangladesh's *haor* basin represents a risky environment and accordingly was among the last regions to be colonised due to the local flood regime, most *haor* land cannot be planted with those rice varieties most familiar to Bengalis. This is because for the local inundation levels cannot even be tolerated by flood-adapted floating paddy varieties. People, however, moved to the *haor* not only because they had no better place to go but also because they must have understood that with some individual and collective effort the area's land and water resources could be used to set up a viable production system.

Whereas in other parts of Bangladesh farmers can grow a variety of crops the whole year around, in the central part of the *haor* basin the only period available for agricultural production is the brief dry winter season and the only crop that can be grown is the *boro* paddy. The conditions under which this single crop is cultivated are stressful and involve a race against the floods. The cultivation season starts at the beginning of November. As soon as the river level is lower than the water level inside the *haor*, farmers need to drain off a portion of the water to start preparing nursery beds for the paddy seedlings. These occupy the highest cultivable land around the homesteads and can be used, after having fulfilled their primary function, to cultivate some winter vegetables and spices.

During the four to five weeks when the seedlings are grown in the nurseries, the farmers are occupied with ploughing their lands. Most farmers cultivate several scattered plots to diminish the risk of crop loss associated with hailstorms, draught, pests and flood.[4]

Chaptir Haor's agriculture could qualify as organic; no fertilisers or pesticides are applied to the paddy nurseries nor to the fields planted with indigenous paddy varieties. Only small amounts of urea are applied to fields planted with HYVs. Even today a major portion of the land is ploughed with the traditional wooden plough pulled by cattle. However, power tillers are seen more frequently here than one would expect from such a remote area. This is related to the shrinking number of cattle resulting from a fodder shortage, which

4. According to the 1996 Agricultural Census farms in Sunamganj district hold an average of 6.8 plots. Even marginal farms (owning less than 0.5 acres of land) hold an average of 2.5 plots, whereas large farms (above 7.5 acres) in average hold an average of 19.7 plots of land.

gets particularly severe during the monsoon period. In fact, whereas other parts of rural Bangladesh provide opportunities for stubble grazing, this is not the case in the *haors*, where paddy fields are flooded immediately after the harvest. The fodder problem has become more severe with the introduction of LLPs, which have allowed to bring under plough cultivation areas of land that used to be available as communal grazing grounds. The number of cattle is thus kept at the minimum required for tilling the land. Many farmers sell their cattle at the end of the agricultural season and buy new cattle the following year. Whereas large farmers generally own sufficient drought animals to plough their land, small farmers who cannot afford to keep cattle have their lands ploughed on a contract basis by holders of power tillers or farmers who own surplus drought animals.

Transplantation of seedlings is done during the first half of December. Whereas drainage at the beginning of the agricultural season is essential to start planting the nurseries and ploughing the fields, farmers are equally concerned to retain sufficient moisture in the soil and to preserve enough water for irrigation in *beels*, *dhobas* and *kharas*. This, as discussed in Chapter 5, cannot be ensured by

Table 7.4: Role of Irrigation in Research Districts

Districts	*Holdings*						
	reporting irrigation (%)	*using surface irrigation*	*using ground-water irrigation*	*using LLPs*	*using STWs*	*using DTWs*	*using tradi-tional irrigation devices*
Sunam-ganj	42.4	40.7	1.7	10.7	1.7	-	33.6
Nawab-ganj	59.6	13.1	64.0	6.1	28.9	20.2	7.9
Chitta-gong	62.4	56.7	7.7	42.6	3.0	3.0	9.8
Patua-khali	2.7	2.7	-	-	-	-	2.7
Bangla-desh	59.5	17.4	49.4	10.8	35.8	7.9	4.5

Note: All these figures are in percentage term.

Source: Bangladesh Agricultural Census (1996)

farmers on an individual basis and constitutes Chaptir Haor's basic collective concern, central to understand the origin of its corporate institutions. *Haor* farmers are aware that indigenous varieties of *boro* benefit greatly from careful water management and irrigation. With the aim of taking advantage of residual soil moisture, as discussed in Chapter 5, they maintain numerous contour dikes (*janghals*). These eliminate or strongly reduce the need for irrigation during the transplanting period. The yields of indigenous varieties of *boro*, however, can be significantly increased through supplementary irrigation during the flowering stage, that is, in March. In this period it may already rain sufficiently to satisfy the crop's basic requirements but not necessarily enough to ensure the highest possible yields. This is one of the reasons why so much land in the *haor* basin was already under irrigation prior to the advent of modern technologies. Even today, as indicated in Table 7.4, out of the 40.7 per cent of the agricultural holdings having access to surface irrigation in Sunamganj District,[5] over 33 per cent rely on traditional irrigation devices. This means that in Sunamganj District 80 per cent of the farms reporting irrigation rely on traditional irrigation systems, which constitute an entirely different situation from the rest of Bangladesh, where irrigated agriculture had a comparatively marginal role prior to the Green Revolution. In Chapai-Nawabganj for example, traditional irrigation systems account for only 13.2 per cent of the irrigated land and, in Chittagong, for 15.7 per cent. Neither modern nor traditional irrigation plays a tangible role in Patuakhali District's agriculture.

The harvest season begins with the local varieties around the middle of April and must be completed by mid-May, when the whole area is again flooded. Local varieties are ready to harvest almost three weeks earlier than HYVs and therefore are less likely to be

---

5. I estimate that district data reflect relatively well the research area's specific situation, except in the case of Chittagong, where the figures are influenced by a number of large-scale surface irrigation systems located elsewhere in the district. In fact, surface irrigation is not that prominent in Banskhali, where we estimate that it may account for not more than 30 per cent of the land under irrigation. Further, in Chaptir Haor, modern irrigation technologies play an even more marginal role than in Sunamganj district as a whole and groundwater irrigation technologies are completely absent.

damaged by flood and hailstorms. Given the fact that the water level in the rivers surrounding Chaptir Haor begins to rise already in March, the harvest is awaited with great anxiety. The local labour force is not sufficient for this task and the *haor* basin has thus historically attracted large numbers of migrating harvest labourers, locally called *dawalla.*[6]

Dawallas are gangs of labourers recruited by a contractor known as *gamasta.* They move to the *haors* by boat from all over Bangladesh.[7] Each boat carries a gang of 12 to 25 *dawallas* who set up temporary shelters in Chaptir Hoar's *patit jami. Dawallas* are only responsible for cutting the paddy and transporting it to the threshing grounds. They are paid a share of the paddy, which they carry back by boat to their native lands after having completed their work. In 1997 Chaptir Haor's *dawallas* were paid 1/7th of the harvest of local varieties and 1/9th of the harvest of *bideshi* (HYV, literally meaning 'foreign') varieties. The earned paddy is divided in equal parts among all gang-members. The *gamasta* receives one share for his role as contractor and supervisor and, provided he was involved, one for cutting the paddy. In a good year, if the harvest can be completed before being damaged by floods or hailstorms, a *dawalla* may earn up to 18 *maunds* of paddy in one month of hard work.[8]

In Chaptir Haor the crops cannot be threshed in the fields or in the homesteads as in other parts of the country. Due to the risk of flood in the fields and the shortage of space in homesteads, the threshing is done on special threshing grounds known as *marer kola,*

---

6. For a detailed analysis of the historical role of *dawallas* in *haor* agriculture see Faraizi (1993) and Islam (1985).
7. Islam (1985) mentions that *dawallas* only come from the neighbouring districts but in Chaptir Haor we found boats coming from as far as Char Faizuddin, one of the most southern points of Bangladesh.
8. In 1997 that amount of paddy had a local market value of almost 4000 Tk. This is a relatively good compensation by Bangladeshi standards, particularly because *dawallas* take their share of paddy back to their home districts, where the paddy price in the same period of the year may be higher. However, it does not keep into account the risks and transport costs. Many *dawallas* in Chaptir Haor complain about being subject to extortion by police and robbers on their way back home, costing them a tangible share of their earnings. For this reason, even if road transport is more costly, some *dawallas* these days choose to rent a truck to transport their paddy back to their homes through the somewhat safer roads.

which are communal high lands located near the villages. Threshings of large farmers are done by their *kamla* (domestic labourers employed on seasonal contract), whereas small farmers rely on daily labourers. Typical of *haor* agriculture is that all operations need to be done under severe time constraints and accordingly even small farms are bound to employ numerous daily labourers. Given the haste with which the harvest has to be completed, a significant amount of grain is left behind in the field. Poor people are allowed to gather this paddy, an activity that mostly involves women, children and older people. This is an important institution which may contribute a tangible portion to a poor household's yearly food requirements. A poor family may collect free up to three *maunds* of paddy, which is sufficient to satisfy their rice-consumption for up to two months.

The conditions under which paddy is cultivated in the *haor* may appear somewhat precarious. However, before the Green Revolution, this crop potentially gave yields higher than any other variety of rice grown in Bangladesh.

If the harvest can be successfully completed before the floods, the month after the harvest is characterised by a sort of collective euphoria and by conspicuous consumption. In this period even the poorest households have enough to eat and are busy preparing various types of *pita*, traditional sweets made of rice flour.

Many traders move to the *haor* in this period to sell their goods (pots, bangles, snacks, toys) from house to house in exchange for paddy. Local fishers may temporarily start a small sweet shop or enter into dry fish business. With these products they primarily target women, the custodians of household granaries to persuade them to sell rice/paddy.

Chaptir Haor's inhabitants, however, are always very conscious that the agricultural season may also have a tragic end; and they don't have to dig too far in the past to recall harvest losses. In 1997 many farmers suffered the consequences of having lost almost 100 per cent of the 1995 crop due to floods, about 50 per cent of 1994 crop due to hailstorms and again about 50 per cent of the 1992 crop due to floods. A vivid description of the vulnerability of *haor* agriculture and some of its socio-economic consequences are given by Islam (1985), who emphasises the point that:

> *Boro*, the haor crop is promptly damaged by lack of rainfall and by the untimely floods. *Boro* is exposed to hailstorm and pestilence. *Boro* crop may be damaged even after harvest if it rains before thrashing and storing the grain. To enjoy the fruits of their labour the peasants must be free from all these evils, such as, too little or too much rainfall, early floods, pestilence, deluge during thrashing and storing and the like. But all these being nature's curses and cannot be regulated by the given technology, *boro* producers face frequent losses of crop. Such loss may be made up quickly by growing supplementary crops in other regions. The *haor* peasant however must wait a full one year for the next harvest and till then they must subsist on borrowings from the *mahajans*.
>
> — Islam 1985:49[9]

This situation stands in contrast to the other research areas, where the failure of one crop may at least partly be compensated by another crop, by alternative sources of income or where a complete crop loss occurs only rarely. For example, an important feature of Beel Singri's farming system is its inhabitants' reliance on the floodplain and the terraced Barind, two distinct physiographic systems. This situation provides a natural insurance against the weather vagaries that so much characterise other parts of Bangladesh and particularly the *haor* basin. Although the situation has changed and floodplain agriculture — thanks to the flood protection works — is safer than it used to be, the description provided by District Gazetteers about agriculture in the area to which Beel Singri belongs is still relevant:

> The topography and the physical condition of the district are such that the whole of it is very unlikely to suffer

9. *Mahajan* is the Bengali term for informal moneylenders who either charge exorbitant rates of interest (in the *haor* basin, around 240 per cent) or take land as a mortgage. Under this system *mahajans* have the usufruct right over the land until the loan is being repaid. Unable to repay their debts, marginal farmers after some time end up being forced to sell their land.

> simultaneously. In the region, mainly comprising of the *Barind*, only one crop, *aman* is grown in a year. This crop is almost entirely dependent upon local rainfall. Nearly the whole of the west of the district (excluding the land near the banks of the river) is low and subject to inundation for at least five months a year. This area is known as *Bhar*. In a very exceptional case, full crops are obtained from both the tracts in the same year, for that would require good rainfall as well as moderate inundation. Crop failures in both the tracts in the same year are equally unlikely. In the year of short rainfall, the crop would be good in the *Bhar*, though those in the *Barind* might fail. If there is high rainfall, the *Barind* will obtain a good harvest, though those in the Bhar may be damaged by the floods. The *Bhar*, moreover, is further protected from famine, by the fact that at least two harvests are obtained there in one year. The variety of crops is such that it is practically impossible that all crops can fail at the same time. Paddy of three kinds (*aman*, *aus* and *boro*) are grown almost universally in addition to various pulses, mustard and sugarcane. Jute is also fairly abundant.
>
> — Siddique 1976:20

Agrarian Structure

With the objective of finding out possible relations between Chaptir Haor's remarkable capacity to overcome collective action problems and its agrarian structure, this section examines the three aspects of this issue, namely: the social distribution of land, the land tenure system and the employment conditions of the landless. These issues will thus be compared with those prevailing in Beel Singri, Banskhali and Galachipa.

DISTRIBUTION OF LAND

Bangladesh's rural economy, as was mentioned in Chapter 3, is characterised by a predominance of small farms, which is coupled with high land concentration. The situation in Chaptir Haor does not basically deviate from the national situation, though, as may be expected from a risk-prone single cropped area, agricultural holdings tend to be larger than the national average.

Chaptir Haor and Beel Singri both have a comparatively low percentage of small farms and their percentages of large farms are, accordingly, well above the national average (see Table 7.5.). A particularly sharp contrast exists, on the other hand, between Chaptir Haor and the densely populated Banskhali area, where already in 1984 over 85 per cent of the cultivated holdings were smaller than 2.5 acres.

Union level data are not yet available for the 1990s. However, district level data, besides indicating a general tendency of farms getting ever smaller, confirm a certain similarity between Chapai-Nawabganj and Sunamganj as well as their sharp contrast to Banskhali.

Table 7.5: Farm size in research areas (1984)

Farm size (in acres)	*Chaptir Haor*		*Beel Singri*		*Banskhali*		*Bangla Desh*
	No.	%	No.	%	No.	%	%
Small farms (0.05 - 2.49)	2,670	55.88	2,407	42.54	21,182	85.26	70.3
Medium farms (2.50-7.49)	1,480	30.98	2,338	41.32	3,335	13.42	24.7
Large farms (7.70 and above)	628	13.14	913	16.14	326	1.31	5.0
Total	**4,778**	**100**	**5,658**	**100**	**24,843**	**100**	**100**

Source: 1984 Agricultural Census (GOB/BBS 1987)

Table 7.6 gives some indication of the actual land concentration, showing that Sunamganj's marginal holdings, although 22 per cent in number, only cover 2.1 per cent of the land, whereas the 5.6 per cent large farms account for 31.3 per cent of the land. This proves that Chaptir Haor, though featuring a certain predominance of medium and large farms (which taken together control almost 75 per cent of the total cultivated land), does not deviate from national figures in terms of land concentration. In fact, as indicated in Table 7.6, land concentration is significantly higher in Patuakhali, where the top 1.6 per cent farms control almost 18 per cent of the land.

With regard to land distribution, a comparative analysis of the four regions does not allow to establish any corelation between the distribution of land and a society's proclivity towards pursuing common goals. This study nevertheless shows that the unequal distribution of land, typical for Bangladesh as a whole, is apparently no impediment for collective action.

LAND TENURE SYSTEMS

An appraisal of the land tenure situation carried out in all research areas reveals that, more than in other regions of Bangladesh, landowners in Chaptir Haor tend to cultivate their land themselves. Though an estimated 20 per cent of the land is cultivated by tenants,

Table 7.6: Farm size in research districts (1994)

Farm size (in acres)	*Sunamganj*		*Nawabganj*		*Chittagong*		*Patuakhali*		*Bangladesh*	
	I*	II**	I	II	I	II	I	II	I	II
Marginal farms (0.05-0.49)	22.0	2.1	28.9	2.3	32.9	6.8	42.2	3.3	27.9	4.3
Small farms (0.50 - 2.49)	45.7	24.4	35.9	19.5	53.3	50.0	21.6	10.2	51.4	36.7
Medium farms (2.50-7.49)	26.5	42.0	29.9	53.2	13.1	38.9	34.5	68.8	18.2	42.4
Large farms (7.70 and above)	5.6	31.3	5.3	24.9	0.7	4.3	1.6	17.7	2.5	16.4
Total	**99.8**	**99.8**	**100**	**99.9**	**100**	**100**	**99.9**	**100**	**100**	**99.8**

* I = percentage of total holdings ** II = percentage of total operated land

Source: Agricultural Census (1996)

this situation is not necessarily associated with a transfer of land from surplus landholders to land-poor tenant farmers. Indeed, Chaptir Haor's case supports Islam's observations about the *haor* basin that "the notion that only poor people lease in land and rich peasants lease out is very wrong" (Islam 1985: 49). As pointed out by this author, rich farmers may lease out distantly located land, which is

Table 7.7: Land tenure situation in research districts (1996)

	*Owner holdings**					*Owner-cum-tenants*					*Owned area self-operated*				
	I**	II	III	IV	V	I	II	III	IV	V	I	II	III	IV	V
Sunamganj	**72.3**	**84.6**	**68.7**	**66.9**	**93.8**	**27.7**	**12.8**	**33.3**	**33.1**	**12.5**	**87.2**	**90.0**	**78.3**	**85.8**	**95.3**
Nawabganj	58.8	90.9	51.8	38.7	70.0	41.2	9.1	46.1	63.3	-	65.9	100	68.3	62.3	74.6
Chittagong	57.4	81.6	46.2	36.4	50.0	41.6	17.3	52.6	63.6	-	63.9	85.2	65.3	59.3	67.6
Patuakhali	61.1	97.4	60.0	18.3	38.6	37.8	1.3	38.0	81.6	72.2	62.6	100	77.2	53.4	69.5
Bangladesh	61.5	76.3	55.0	58.0	71.6	37.5	21.9	44.0	41.5	33.3	78.4	80.3	73.4	79.9	88.0

* The category 'owner holdings' refers to farmers cultivating only their own land. Because of its marginal importance, the category 'pure tenants' is not reported here.

** (I) All holdings; (II) Marginal holdings (0.05-0.49 acres); (III) Small holdings (0.50-2.49); (IV) Medium holdings (2.50-7.49); (V) Large holdings (7.70 and above)

Source: Agricultural Census (1996)

more difficult for them to manage and lease in more conveniently located plots. Medium farmers, who own land below their management capacity, are most inclined towards renting in, whereas marginal landholders, for whom the management of the farm often becomes too costly, are inclined towards renting out. In addition to the land leased out by people who live in Chaptir Haor, some land for lease is available from migrated households, who tend to consolidate their socio-economic status in their native home by buying up more land. These lands often are leased in by local relatives, though they may themselves be large farmers.

District level statistics support these observations. As shown in Table 7.7, an average 87.2 per cent of the land is cultivated by owners. The table also confirms that all categories of landholders are involved in leasing land. This tendency cannot necessarily be attributed to the strongly agrarian character of Chaptir Haor's economy. As was indicated earlier, agriculture also dominates the economic structure of Beel Singri and Galachipa, two areas showing a pattern of leasing out land similar to the one found in the more urbanised district of Chittagong. An important reason why most farmers in Chaptir Haor prefer to keep direct control over their land is that most people in this region do not hold land outside their own *haor*, which means that all their land is likely to be located within their reach. This is clearly not the case in Beel Singri and Galachipa. In Beel Singri, for example, in addition to land in the floodplain, large farmers own large estates in the dry and single-cropped Barind tract, located far from their residence. Likewise, Galachipa's powerful elite controls, though not necessarily legally, more land in the surrounding *chars*, than what they own in Galachipa. These scattered holdings make it virtually impossible for large landlords to cultivate all their land by themselves.

Only in Chittagong, where a considerable amount of land is leased out despite the comparatively small size of the holdings, this tendency may be attributed to a more diversified economic structure, allowing landowners to take advantage of non-agricultural employment opportunities.

Tenancy conditions in Chaptir Haor, as indicated in Table 7.8, do not differ dramatically from those prevailing in other areas.

Sharecropping, locally called *haja,* is the most common type of tenancy contract. In Chaptir Haor, under this type of contract the landlord receives 50 per cent of the harvest without participating in the costs of production. This is the most common type of sharecropping contract all over Bangladesh in case of low-input crops. The conditions change when the cultivated crop requires modern and capital-intensive inputs, in which case, as indicated by Banskhali's and Beel Singri's situation, the owner manages some of the costs of irrigation and agro-chemicals. The sharecropping conditions in Galachipa are outstandingly bad. In this area tenants have to pay 2/3 of the harvest to the landlord, without the latter contributing anything to the costs of production. As mentioned earlier, in Galachipa, rich farmers may often control more land in the surrounding *char*s, than in the polder. By offering comparatively favourable conditions in the *char*s they provide an incentive to tenants to cultivate land far away from their families, where, due to the poor development of infrastructure and services and to the risk of cyclones, living conditions are more precarious.

Another type of tenancy contract found in Chaptir Haor is locally called *chukti* and refers to the payment of a land-rent corresponding to a fixed amount of paddy. In 1997 the prevailing rent under a *chukti* contract was of 5-6 *maund*s of paddy per *keyar* of land, which corresponds to about 42 per cent of the harvest. Under a *chukti* contract the farmer bears the full risk of the crop being partially or completely damaged. Accordingly, as may be expected, there is a clear preference among tenants for *haja* contracts. In Beel Singri, this fixed lease in kind is called *puran*, meaning 'old system' and represents the most common tenure-contract for the traditional *aus* and *aman* crops. Though this system is named differently in Beel Singri, the terms and conditions are about the same as those in Chaptir Haor, to the extent that in Beel Singri the land rent corresponds to about 40 per cent of the harvest. Another relatively common tenure contract found in Chaptir Haor, locally called *rangjama*, consists of a fixed rent in cash. The equivalent type of contract in Beel Singri is called *khaikalashi*, with terms and conditions that are very similar to those prevailing in Chaptir Haor. The tenure conditions prevailing in Banskhali reflect a more diversified use of land. Sharecropping, for practical reasons, prevails for paddy production but is not applicable to other land uses such as horticulture, salt and shrimp production. Accordingly, fixed land rents in cash are far more common in this area than those characterised by a more homogeneous, paddy-centred, production system.

As indicated in Table 7.8, there is strong variation across the four regions with regard to the mortgaging conditions. This may be explained by the fact that mortgaging conditions are related to the market price of land, which is much below the average in the single cropped *haor* area and much above the national average in densely populated Banskhali. Summing up the land tenure situation, the main difference between Chaptir Haor and other parts of Bangladesh is that — to a far greater extent than in other areas — land there is cultivated by the owner. This reality may be explained by the local economic structure, which provides almost no employment opportunities other than agriculture and fishery.

LANDLESSNESS

According to the 1991 Population Census 50.8 per cent of the households living in Chaptir Haor do not own any agricultural land. The percentage of landlessness is about 5 per cent above the national average, revealing a particularly dramatic situation, considering that Chaptir Haor's landless have few employment opportunities beyond agriculture. During the monsoon, the limited employment opportunities are partly offset by the free access to communal fisheries and other aquatic resources (e.g., wild vegetables). Without access to these free resources it would be difficult for Chaptir Haor's landless to survive until the next season.

Islam (1985) maintains that the high percentage of landlessness that characterises the *haor* basin is closely associated with the vulnerability of local agriculture. Crop failure systematically forces small farmers to mortgage and eventually sell off their land. Though individual cases may support this interpretation, this argument is implicitly based on the assumption that once upon a time Bengal constituted a more or less undifferentiated peasant society, which was hardly the case now. A general correlation between landlessness

Table 7.8: Tenure systems in research areas

Research areas	*Tenure contract*		*Conditions*
	Local name	*English*	
Chaptir Haor	*haja*	sharecropping	Harvest, including by-products, is divided between landlord and tenant in equal parts. No participation of the landlord in production costs.
	chukti	seasonal lease, fixed rent in kind	5-6 *maunds* paddy per *keyar*
	rangjama	seasonal lease, fixed rent in cash	800-1,000 Tk per *keyar*
	bondok	mortgage	3,000-3,500 Tk per *keyar*.
			(Land is returned to owner upon repayment of the loan)
Beel Singri	*adi*	sharecropping	Same like in Chaptir Haor but landlord pays 50% of fertilisers and 50% of irrigation costs. Prevails during *boro* season
	puran	fixed lease in kind (literally meaning "old system")	3-5 *maunds* paddy per *bigha*. Prevails for *aus* and *aman* crops
	khaikalashi	fixed lease in cash	600-1,000 Tk per *bigha*
	khajna kaura	mortgage	8,000-10,000 Tk per *bigha*. (Land is returned upon repayment of loan)
Banskhali	*baga*	sharecropping	Same as in Beel Singri
	lagit	fixed lease in kind	13 *maunds* per acre
	agu	fixed lease in cash	3,000 Tk per *kani* for paddy land

			10,000 Tk per *kani* for salt production 6,000 Tk per *kani* for vegetables
	bhanda	mortgage	15,000 Tk per *kani*
Galachipa	*bhaga*	sharecropping	*aman* in polder: 2/3rd of the harvest goes to the landlord
			aman in *char*: 50% of the harvest goes to the landlord
	khaikalashi	7- year contract in cash	*aus*: 50% of the harvest goes to the landlord 7,000 Tk per acre
	kabla	mortgage	10,000 Tk per acre (Land returned up on repayment of loan)

Note: 1 acre = 100 decimals; 1 *keyar* = 30 decimals; 1 *kani* = 40 decimals; 1 *bigha* = 33 decimals; 1 *maund* = 37.3 kg

Source: Indepth individual and group interview with key informants

and vulnerability of the production system is also contradicted by the case of Beel Singri where the percentage of landless households is even higher than in Chaptir Haor, in spite of the multiple employment opportunities and agro-ecological diversity.

Indeed, few among Chaptir Haor's currently landless households ever owned any land. Many, though having lived in the *haor* basin for several generations, are historically landless *kamla* (labourers). Others have lost their land in other areas of Bangladesh due to river erosion or other reasons and migrated to Chaptir Haor to escape from penury in their native areas. Indeed, the labour demand in Chaptir Haor, though interrupted by prolonged slack season, is high; and labourers are comparatively better paid than might be expected from such a remote area.

Table 7.9: Landlessness in research areas

Research areas	*Households*	*Owning agricultural land*	*Landless*	*Landless (%)*
Chaptir Haor	7,080	3,497	3,583	50.61
Beel Singri	12,839	4,582	8,257	64,31
Banskhali	44,236	22,879	21,357	48,28
Bangladesh	n.a	n.a	n.a	44.49

Source: 1991 Population Census (GOB/BBS 1994)

Many among Chaptir Haor's landless find employment as seasonal domestic labourers. In fact, as was mentioned earlier, the small farm relying on family labour hardly exists in the *haor* basin. Due to the strict time constraints in which rice has to be grown, even small farmers rely on hired labour. Among medium and large farmers it is very common to hire several labourers on a seasonal basis. These domestic labourers live and eat with the household of the landlord and are recruited in November, the beginning of the agricultural season, for a period of six months. In 1997 they were given three meals per day and paid 25-30 *maunds* of paddy plus one *lungi*, one *gamcha* and one *ganji*.[10] Some large farmers may hire a few domestic labourers

10. Traditional male clothing items: the wrapped ankle-length garment (*lungi*), a towel for covering head or shoulders (*gamcha*) and a undershirt (*ganji*).

year-round. During the slack season they are paid in cash, between 500 and 600 Tk per month plus food and lodging. This is not much, but better than nothing, considering that the alternative could well be starvation. Large farmers may also employ young boys to look after their cattle on a seasonal or year-round basis. Depending on their age, they are paid 10-15 *maunds* of paddy plus meals and clothes. For poor families, sending their children to live and work outside their home is a strategy to reduce expenditures, to increase income and to obtain patronage from better-off families.

The demand for female labour is particularly high during the harvesting season. Women are responsible for cleaning, drying and winnowing the paddy. During the harvesting season female labourers are paid for a one-month period, in addition to their meals, with 4-6 *maunds* paddy, a *sari* and a bar of soap. This is significantly more than what they can earn during the rest of the year, when monthly female wages range between 200 and 300 Tk.

The agricultural daily wages prevailing in Chaptir Haor compare relatively well to those of Beel Singri and Galachipa, where daily wages, except during brief peak demands, are only 25-30 Tk per day plus meals. In Beel Singri daily wages reach 50 Tk per day (plus meals) only during the two weeks of *aman* paddy transplantation, whereas in Galachipa they reach this level only during the *aman* harvesting season. Agricultural daily wages are somewhat higher in Banskhali but generally they do not include any meals. In this area daily wages vary between 70-80 Tk per day without meals. If meals are provided, the daily wage is only 50 Tk, which indicates that three meals have a value of about 20 Tk.

The seasonal fluctuations and regional differences in daily wages reflect local variations in labour supply and demand. In Chaptir Haor the demand for agricultural labour is high but concentrated in a period of only six months. To ensure the availability of labour when required, the conditions are bound to improve at least partially for the long slack season. In Beel Singri and Galachipa wages are low but the period during which landless households cannot find any employment lasts for only about three months. A comparison of the labour situation in the four regions does not indicate that some regions in Bangladesh are more equitable than others. All over Bangladesh, the conditions of landless labourers are precarious, which are clearly reflected in official statistics showing that between 45-53 per cent of the country's population live in more or less poor conditions while 23-35 per cent population live in extremely poor conditions (Rahman 1998: 5). Nevertheless there is a qualitative difference between Chaptir Haor and the other regions. First of all, there is a closer, more intimate contact between labourers and their employers. Secondly, there is a marked tendency to remunerate labour in kind. These conditions, may contribute to the fact that also landless people in Chaptir Haor strongly identify with the place and its resources and actively support collective initiatives in water management, particularly in case of emergencies.

Haor Fishery

The fishing sector and its relation to agriculture is of critical importance to understanding collective action in Chaptir Haor. Fishing practices, though not in conflict with agriculture, constitute a potential threat to the *haor*'s single crop. Contrary to other parts of Bangladesh, where the conflict between agriculture and fishery is almost proverbial, this is not the case in the *haor* area, where there is a considerable interdependence between these two sectors. In fact, at the end of the monsoon, though for different reasons, both farmers and fishers are interested in drainage. Drainage through cuts in the river levees or embankments and removal of dams is required by farmers to start tilling their lands as early as possible. For fishers, drainage constitutes a traditional fishing technique whereby fish carried by the current are caught in nets installed at the drainage points. Drainage also contributes to concentrate on the remaining fish in a smaller area. But farming and fishing activities need to be coordinated to provide the best possible conditions for the *haor*'s vulnerable paddy crop.

There are no written rules and no formal authorities responsible for the overall management of Chaptir Haor's water resources but there is a general consensus that agriculture gets priority over fishery. For example, it is known to everybody that the leaseholders of *beel*s and *dhoba*s are not allowed draining these surface water bodies without farmers' agreement. It is not allowed to cut the embankment for final fishing before *Phalgun* 15th (beginning of March). Before cutting the embankment and draining any amount of water, leaseholders have to seek permission from the *matabars* of those villages whose lands are affected by the cuts. After completing drainage, the leaseholders have to close the embankment cuts. Traditionally they had to deposit an agreed-upon amount of money with the *matabars*, as a commitment towards this responsibility.

Another clear relation between Chaptir Haor's fisheries and its collective water management practices relates to the local use of *dhobas*. As was mentioned earlier, these common properties are instrumental to generate village funds. Given the importance of these funds to sustain the *haor*'s indigenous infrastructure, if villages would not use *dhobas* to generate them, they would most likely have found another solution, such as organising more taxation or collecting taxes from the villagers. Thus, though collective action in Chaptir Haor may not be explained by the fact that the area owns more communal resources than other regions, the existence of common resources certainly represents a supporting condition.

A comparison between the role and organisation of fishery between the four research areas points at a number of critical differences, which are important to understand why Chaptir Haor has been more successful in overcoming its collective action-problem. If we compare Chaptir Haor with Beel Singri, for example, it may be noticed that in Beel Singri, as long as there was no modern hydraulic infrastructure, there was no interdependence between agriculture and fishery. Year-round cropping patterns were adapted to the flood regime and indigenous fishing practices did not have any effect on agriculture. The situation changed with the construction of modern sluices allowing for regulation of water levels in their areas of influence. These have led to a severe conflict of interest between agriculture and fishery. Given the weak socio-economic status of fishers, the situation did not evolve into an overt conflict or negotiation among farmers and fishers but simply resulted in a rapid decline of local fish resources and in the migration of traditional fishing communities to neighbouring India. This explains why today, as indicated in Table 7.10, fisheries in Beel Singri have started playing comparatively marginal role.

Table 7.10: Households involved in fishing in research areas (in %)

	Chaptir Haor	*Beel Singri*	*Banskhali*	*Galachipa*
Households involved in fishing	*72*	*38*	*61*	*65*
(a) for subsistence only	78	68	81	46
(b) for sale only	3	6	12	40
(c) for sale and subsistence	19	26	7	14

Source: Soussan *et al.* (1998: 26)

As in Beel Singri, also in Banskhali and Galachipa traditionally there was no particular interdependence between the organisation of agricultural production and fishery. Accordingly, related practices did not require regulation by public rules. In these two regions, however, the importance of fish culture is currently increasing. Conflicts of interest between agriculture and fishery are gradually emerging in those places where fishery schemes are associated with private individuals' encroachment in common properties, thereby reducing the water available for agriculture or causing drainage congestion. Whether in these areas people will be able to overcome these problems by reaching a consensus on some basic norms and by imposing restrictions on individuals remains to be seen. However, given the fact that the groups affected by these relatively small fishery schemes are far more heterogeneous in terms of water management interests from those in Chaptir Haor, it is not likely to be the case.

The Role of the State

As was mentioned in Chapter 2, state attempts to 'develop' rural societies and govern their natural resource systems affect people's inclinations towards operating collectively at the local level. In Bangladesh, at least over the last three decades, the state has had a relatively strong presence in rural areas. Particularly since the emergence of Bangladesh as an independent nation, each successive government has given much importance to the economic and political integration of its rural areas. Nevertheless, due to the unequal development potential of different regions and to the geographic distance that separates remote areas from the centre, state interventions in rural Bangladesh have not been very homogeneous neither in their intentions nor in their results.

The *haor* basin counts as one of the most remote and backward regions of Bangladesh. Its natural environment is generally perceived as a major constraint, providing little if any scope for development. Accordingly, the *haor* basin has been rather neglected by governmental development agencies, whose presence beyond Sunamganj, the district headquarters, can hardly be felt. Due to the uncomfortable living conditions, many government officials perceive being stationed in this area as a sort of punishment, a sense which does not necessarily enhance their professional commitment as civil servants.

Although *haor* agriculture is contingent upon flood control, drainage and irrigation, the only water management issue that until recently has received official attention from water development authorities is the area's proclivity to flash floods. In fact, the BWDB arrived in this region only in the late 1970s and with a much lower profile than in other areas. BWDB's endeavour to develop the basin's hydraulic infrastructure was carried out within the framework of the internationally funded "Haor Development Programme" and primarily consisted in the construction of 'submersible embankments.' In Chaptir Haor, the state provision of submersible embankments did not imply much else than piling up earth on a stretch of 42.7 km of the natural river levees, a work that prior to state intervention was carried out to a large extent by the people themselves. Most works were done under the World Food Programme's "Food for Work Programme." This means that the associated works were paid in wheat and that the whole project, though carried out under the auspices of the BWDB, had a certain relief connotation. The submersible embankments only aim at increasing the probability that farmers can harvest their indigenous paddy before the entire area is submerged; and accordingly, though they have somewhat reduced the risk of crop loss, it did not result in any structural change for the local agriculture.

Ever since the BWDB has raised these embankments, it has also taken over periodic maintenance responsibilities. These works need to be carried out every year after the monsoon water recedes. The volume of work carried out each year is not based on an assessment of the local conditions. It depends on the wheat made available by the WFP or rather on the share that the BWDB regional office manages to secure for this region. Until recently, no institutionalised effort had ever been made to plan the works to be done by keeping into account locally felt needs and priorities. In the mid-1990s the BWDB came under international development agencies' pressure to set-up beneficiaries' organisations to enhance participatory water management, but hardly made any effort in this direction. Nevertheless, informal interaction between BWDB officials and the local elite takes place rather regularly, even if their demands are not necessarily taken into consideration.

Considering that the BWDB did not do anything other than improving the viability of some of Chaptir Haor's indigenous infrastructure, it is not surprising that most people basically appreciate the infrastructure provided by this agency. However, many people believe that since the embankments were raised the water level in the rivers surrounding Chaptir Haor has risen almost equally and accordingly this has neutralised the effect of this improvement. Another reason why many people in Chaptir Haor are not overly enthusiastic about the BWDB is their suspicion that the BWDB has far more means at its disposal to maintain Chaptir Haor's embankments than what they actually spend. Indeed, there are persistent rumours that only minor portions of FFW wheat are effectively spent on maintenance. Local people are often grumbling about BWDB's failure to maintain adequately the *haor*'s embankment and do not restrain from judging its engineers as unresponsive and corrupt government officials.

Chaptir Haor's farmers feel that they cannot trust this agency and accordingly have never lowered their guard. With regard to the embankment's need for periodic maintenance (in particular in those portions damaged by the passage of boats or that are cut each year to facilitate drainage), they are ready to intervene in case the BWDB fails to ensure an adequate status of vital portions of the embankment, that is to comply with what formally corresponds to their obligation. The BWDB, however, has neither transformed Chaptir Haor's indigenous agriculture nor has it reduced people's vulnerability and interdependence. Nevertheless, the fact that the BWDB is an ever present subject of discussion among farmers indicates the importance attributed to this agency's mandate.

Besides the BWDB, no other development agency has had a tangible impact on Chaptir Haor's livelihood system. Due to the *haor's* unreachable aquifers and the fact that the *haor's* indigenous *boro* varieties were not believed by agricultural scientists to require any substantive amount of irrigation, BADC has never targeted this region. LLPs, which in these days are relatively common in Chaptir Haor and which have allowed for bringing about 13 per cent of its agricultural land under cultivation with *boro* HYVs, have been purchased by private farmers without any government incentives.

People in Chaptir Haor are aware of being neglected by the government and are convinced that the state, with its property claims over the *haor*'s valuable *beel*s, is gaining from the *haor*s much more than it has ever invested for the area's development. In this domain, the *haor* people have perhaps learned from the government that fish may be considered a valuable source of revenue and take advantage of their remoteness by not surrendering their *dhoba*s to state control.

Chaptir Haor's remoteness and the limited role of the state in the provision of hydraulic infrastructure is important to understanding the perpetual need for collective action. If the BWDB were more reliable, had a more comprehensive perception of the *haor* basin's water management requirements and had more means to sustain the local infrastructure, people would gladly hand over this burden to the state. Indeed, the fact that Chaptir Haor's inhabitants solve most of their water management problems on their own does not mean that they would not welcome external assistance or that they do not consider it the government's responsibility to maintain at least the infrastructure for which it claims legal ownership. In fact, ever since the BWDB has been present in the Chaptir Haor, its village *matabar*s, before allocating any effort and resources to the annual maintenance of its embankments, first try to persuade the BWDB to take over responsibility for these works. To do so, they do not hesitate to present themselves as humble and impotent victims, too poor and powerless to solve their problems themselves. Only when they realise that they will not succeed, they mobilise their own resources. This shows that local dependence from external aid is not as strong and irreversible as is often assumed and that the relation between the local level and the state, even without any formal mechanisms to ensure participation, is rather dynamic and interactive.

Compared to the other research areas, the role of the state in Chaptir Haor has been far more marginal. In Galachipa and Banskhali, for example, BWDB's coastal embankments had a significant impact on local people's livelihood system. By reducing the intrusion of saline water to a minimum, it became possible to intensify agricultural production and to bring under plough-cultivation large portions of land that were previously not suitable for cultivation. In Banskhali and Beel Singri the state also had an important role in the dissemination of groundwater technologies and other modern agricultural inputs. Nevertheless, a stronger presence of the state in these areas does not explain why there are significantly less collective actions in water management than in Chaptir Haor, for there is no reason to assume that the BWDB or other government agencies in these regions have systematically replaced endogenous initiatives. Quite to the contrary, as was shown in Chapter 5, many local water management initiatives in Beel Singri, Banskhali and Galachipa are of recent date and were directly or indirectly triggered by state-provided infrastructure and associated agricultural development programmes.

This study has shown that state interventions at the local level have a strong impact on people's individual and collective livelihood strategies but do not explain why certain areas, such as, for example, the *haor* basin, have developed stronger institutional mechanisms than others to overcome collective action-problems.

Conclusions

This chapter is aimed at giving some possible explanation as to why people in the *haor* basin are better organised to mobilise human and material resources for collective purposes than those in other regions. A detailed description of the *haor* socio-economic and agro-ecological context and a comparison with the other research areas showed that among the four regions covered by this research, there are some similarities as well as some clear differences.

An analysis of the socio-economic context revealed that Chaptir Haor's people exceptional capacity to secure collective interests in water resource management cannot be explained in terms of regional socio-economic differences; Chaptir Haor's economy is much more agrarian than Banskhali's, but not more than Beel Singri's and Galachipa's. Chaptir Haor shares with Beel Singri — the area with the lowest number of endogenous initiatives — a distribution of land more unequal than the national average. The much more even distribution of land characterising, for example, the agrarian structure of the Banskhali research area, further indicates that there is no correlation between the degree of a group's socio-economic differentiation and its capacity to pursue collective goals.

With regard to land tenancy, Chaptir Haor's conditions are neither better nor worse than those prevailing in other areas, particularly if the comparatively low productivity and high risks associated to *haor* agriculture are kept into account. Tenants and landless labourers in Chaptir Haor do not enjoy conditions that hint at a higher degree of community which could explain their readiness to contribute with voluntary labour to initiatives benefiting primarily the land-holding rural elite. Thus, the nature of Bangladesh's agrarian structure with all its micro-level variations, contrary to what has been argued by Boyce (1987) neither explains why people in some regions succeed in organising collectively nor why in other regions they do not. It rather supports, as was recently found by Bardhan and Johnson (2000) and Vedeld (2000), that social inequality is not necessarily an impediment for collective action.

A comparative analysis of the four regions indicates that Chaptir Haor may be considered as a relatively neglected area, with regard to state interventions. In particular in the domains of agriculture and water management, the area has been excluded from national development strategies. The government's weak presence in the Haor Basin may explain why people in Chaptir Haor continue to depend to a considerable extent on their own initiatives and resources to protect their production system. It also explains why they have been successful in keeping local control over some revenue generating communal resources of which the government could possibly claim ownership. However, as indicated by many of the endogenous water management practices in the other three regions, government interventions do not necessarily replace local initiatives; quite to the contrary, numerous cases in Beel Singri, Banskhali and Galachipa, indicate that governmental projects may indeed indirectly even trigger them. It was further argued that there is no evidence that in the other research areas, prior to state interventions in the water sector, there used to be the type of collective water management practices and related institutions found in Chaptir Haor. The relatively weak presence of the state in the *haor* basin is a consequence of the limited applicability of the prevailing development strategies and related technologies to the *haor* natural environment: its unreachable aquifers do not allow the introduction of groundwater irrigation that would have reduced farmers' interdependence on communal surface water irrigation systems. Its flood regimes could not be influenced by state-provided hydraulic infrastructure; and neither was it possible to promote in the *haor* area the widespread adoption of HYV and related modern agricultural inputs.

What differentiates Chaptir Haor more than anything else from other parts of Bangladesh is neither its social structure nor its relation to the state but its fascinating natural environment. People in Chaptir Haor are organised because their livelihoods depend on collective action and because, in case they fail to seal off the boundaries of their *haor* and thus to protect it from floods, together they all lose their only annual crop. Indeed, farmers in Chaptir Haor, regardless of their power status and the size of their property, depend on the same crop, for the environmental conditions do not allow any diversification. To protect this — for Bangladeshi standards — exceptionally undiversified production system, people have no choice but to cooperate.

The case of Chaptir Haor and its comparison to other parts of Bangladesh supports Robert Wade's theory that the emergence of corporate institutions is primarily explained by ecological factors. In Wade's south Indian villages ecological factors determined a high risk of conflict and scarcity of water for irrigation. Likewise, in Chaptir Haor the high risk of crop loss may be considered the prime reason behind much collective action. Whereas water in Chaptir Haor compared to south India is certainly not scarce, only through careful management can it meet the requirements of the local crop. This is not the case in other areas, where the traditional production system is based on rain-fed paddy, where traditional cropping patterns provide a natural insurance against floods, and where nowadays — to meet the water requirements of the newly introduced winter paddy — groundwater irrigation provides the opportunity to pursue individual strategies, hence to 'exit' from collective action.

8

General Conclusions

Determinants of Collective Action: Group Attributes *vs.* Contextual Factors

THIS study shows that people in rural Bangladesh, despite poverty and inequality, have the capacity to organise and mobilise resources to make their environment safer and more productive. Through construction, operation and maintenance of water management structures, people increase the area of land under cultivation, reduce drainage congestion, protect land and properties from floods, secure water for domestic uses, and so on. Most of these initiatives are taken endogenously, that is, without material or organisational support from external agencies.

These findings are somewhat counterintuitive. For over two decades social scientists have explained agricultural stagnation and the failure of several rural development projects in Bangladesh as resulting from the prevailing unequal distribution of power and productive resources, which is assumed to hinder cooperation. Indeed, much research on Bangladesh shows that rural elite groups often have been able to hijack farmers' organisations and appropriate for their own benefit public development resources channelled through them. Accordingly, most scholars have concluded that cooperation among people divided by socio-economic inequality and factional conflicts is unlikely to occur. This book does not intend to glorify people's capacity to cooperate. Rather, it aims at understanding the conditions under which groups of people are likely to participate in the provision or maintenance of a collective good.

Collective action theorists have sought answers to this general question by focusing primarily on group attributes, particularly on *size* and *heterogeneity*. The present study shows that size does not necessarily prevent a group from organising but it does have an impact on *how* resources necessary for the provision of a collective good are mobilised. Most initiatives presented in this study are undertaken by small groups, indicating that it may be easier for a small group to organise and further a common goal than it is for a large one. In all regions covered by this study, none the less, we find instances of relatively large groups organising collectively in order to enhance or protect agricultural production.

Small and large groups differ, however, in their strategies to mobilise required resources, as Olson (1965) also has observed. In small groups, contributions among members are generally collected in proportion to individual benefit, whereas in large groups resources are mobilised by appealing to people's solidarity, by generating funds through auctions of common properties or by collecting 'taxes' on activities such as gambling. These differences in patterns of resource mobilisation cannot be attributed to differences in group-size alone. They also are explained by the nature of the good.

To make this point clear, this study has differentiated between 'collective goods' and 'public goods.' Collective goods yield quantifiable returns to each group-member, who may be easily excluded from any benefit if they fail to contribute. In the case of public goods it is difficult to distribute the burden among group-members in proportion to their individual benefit and also to exclude free-riders from using the good. This explains why Chaptir Haor's inhabitants, all of whose livelihoods are contingent upon the provision and maintenance of public water management systems, were bound to develop some strategies to mobilise resources, which would assure their availability on a sustained basis. The capacity of Chaptir Haor's 35 villages, having a total population of almost 50,000 people and scattered over an area of 50 km^2, to organise together proves that the large size of a group is not necessarily an impediment for collective action.

In all the cases presented in this study-group-action was organised by a few leading individuals who invested more time in the initiative than the average group-member, took upon themselves the organisational costs and often also contributed a disproportional share of the material and financial resources. This supports Marwell and Oliver's (1993) thesis that collective action primarily depends on a critical mass of highly interested people.

The influence of group heterogeneity on collective action is perhaps one of the most interesting and controversial issues in collective action theory. Most differences of opinion, however, are to be associated to the different meanings given to this concept. This study shows that, depending on which dimensions of group heterogeneity are being considered, it can either have a positive or a negative impact on collective action. If heterogeneity is defined, as it is by Olson (1965), as the differential value placed by individuals on a collective good, it generally has a positive impact on collective action. For example, in many cases covered by this study, when a group of farmers needs a common irrigation system, large farmers may be more interested in the system being provided than small ones and accordingly be willing to take upon themselves the transaction costs incurred in organising the group and mobilising resources. This explains why in many water resource management situations, rich and powerful elite groups have had a positive role in the provision of water management systems, and social inequality was no impediment for cooperation. In such cases, occupational homogeneity, in terms of all group-members being farmers and thus agreeing on a specific use of surface water resources, was more important to understand the outcome of collective action than socio-economic homogeneity.

Nevertheless, occupational homogeneity by itself is no guarantee for the resolution of common problems. The potentially positive effect of occupational homogeneity on collective action may be obstructed by differences in the location of people's land resources. This is the case in Beel Singri, where lowland farmers' interests conflict with those of highland farmers. The case of Beel Singri also points at the importance of conflicting interests, among elite groups, that is, among people with a similar socio-economic status. In fact, the fuzzy operation of BWDB's drainage regulators was the result of interest heterogeneities among equally powerful people.

Problems caused by locational heterogeneity of endowments have gained much attention in irrigation systems, where it is well known that head-end farmers and tail-end farmers have different incentives to cooperate. The situation, however, is even more complex in the case of floodplain water management. Whereas in irrigation systems group-members basically consist of farmers wanting irrigation, in the case of Beel Singri there are conflicting interests between farmers and fishers, between lowland farmers and highland farmers and between farmers having access to alternative sources of irrigation water and those without access. The situation is further complicated by the option of groundwater irrigation being available to some but not all farmers. Under such conditions it is no surprise that Beel Singri's inhabitants are generally far less successful in finding equitable solutions to their common problems than are those of Chaptir Haor.

The above considerations about the different types of heterogeneity in Chaptir Haor and Beel Singri show that neither an analysis of group attributes nor of the nature of the good could explain why, in terms of their capacity to organise collectively, there are such remarkable qualitative and quantitative differences between the four regions. As was shown in Chapter 5, by far the highest number of local initiatives are found in Chaptir Haor. This area's collective actions are distinctive not only in terms of numbers of people and resources involved but also with regard to the historical role of endogenous hydraulic systems supporting people's livelihood, their degree of institutionalisation and the general social consensus backing them.

In order to explain these differences, this book proposes a systematic analysis of contextual variables, defined as external or internal factors that may influence the demand of a public good and the willingness of people to contribute to its supply. Accordingly, the comparative contextual analysis proposed in Chapter 7, has focused on the research areas' natural environments, their socio-economic conditions and the role of the state.

It was shown that from a socio-cultural and economic point of view, there is no significant difference between the four research areas or at least that there is no corelation between these differences and the inclination of an area's population to engage in collective actions. In fact, Banskhali, the only region with an especially diversified economy, shows no tendency toward a lower capacity to pursue common goals. In particular, it was found that the research areas do not differ much in their "degree of community", which according to Singleton and Taylor (1992), should explain why some groups are more successful than others. In fact, among the four regions, Chaptir Haor is the most recently settled, with many among its inhabitants having lived there for less than three generations. Further, Chaptir Haor has international migration rates above the national average, and — following Singleton and Taylor's thesis — should thus have a lower degree of community than the much earlier settled and more stable population of Beel Singri. This shows that many contributions to collective action theory are unable to explain qualitative and quantitative variations in collective action across socio-culturally homogeneous regions.

An analysis of the role of the state, focusing in particular on public interventions in the agricultural and water development sector, shows that Chaptir Haor has been comparatively neglected by governmental development assistance. This explains to some extent why people in Chaptir Haor have had to rely on their own resources to provide themselves with many of the public goods they need. Of course, it cannot be denied that the BWDB could and already did take over some of the water management responsibilities that used to be assumed endogenously. Nevertheless, as was pointed out in Chapter 7, there are two reasons why state interventions cannot explain why some regions are better organised than others. First, there are no indications that in the other three research areas, prior to state interventions, endogenous initiatives in water resource management ever had the importance they still have today in Chaptir Haor. Secondly, as indicated by many initiatives in Banskhali, Beel Singri and Galachipa, state intervention may not necessarily reduce the need for endogenous initiatives. State intervention may actually create new interdependencies and opportunities, thus enhancing the need to cooperate. Modern, state-provided flood control and drainage systems have in some regions significantly increased people's interdependence and thus their mutual vulnerability. On the other hand, the introduction of groundwater irrigation pumps has provided, to some areas and some people, new opportunities that reduce their need to cooperate.

Whether and how effectively new technologies can be introduced ultimately depends on environmental factors. Chaptir Haor's singularity is not related to socio-cultural factors; nor can it be explained by the limited role of state agencies. But the area is certainly unique from an environmental point of view. As was shown in Chapter 7, most *haor* land remains submerged in deep water for over six months a year; and only through immense collective efforts is it possible to cultivate a single crop of paddy during the short dry season. The high risk of loosing the crop to flash floods, which occur frequently enough to require collective disaster preparedness, and to which all people, regardless of their endowments, are exposed, makes people mutually vulnerable and more interdependent than in any other area of Bangladesh.

If Chaptir Haor is unique, what do the other three regions have in common? At first glance, one might say, not much. Beel Singri is located in an economically backward interior region. It is relatively dry but, thanks to its topography, the availability of groundwater and the possibility to store surface water, farmers can cultivate many different types of crops throughout the year. Banskhali is a relatively wealthy coastal area with a diversified economic structure. Good access to urban areas offers many people some alternative non-agricultural employment opportunities; and there are possibilities for horticulture. Galachipa is a remote coastal area that in many respects has maintained the socio-cultural features of a *char* society. Its local elite is more interested in grabbing new lands emerging between the river beds than in optimising use of the land they already control. These three regions have in common, first of all, patterns of human adaptation that are to a large extent individual. Secondly, compared to Chaptir Haor, they also have in common a much lower level of risk; nowadays, thanks to state-provided flood control systems, their agriculture is no longer regularly threatened by floods. Thirdly, they all offer some exit strategies. In Beel Singri and Banskhali one exit strategy adopted by many farmers is turning to private groundwater irrigation. By doing so, they can avoid the organisational difficulties associated to surface water irrigation. In Galachipa, many people have the option of turning their backs on the local land resources and increasing control over *char* lands. Such exit strategies, as pointed out by Hardin (1982), dissipate prospects for collective action.

This study has shown, in conclusion, that collective action in Bangladesh, though with significant regional variations, plays an important role in rural people's livelihood and that people — even if divided by many differences — in some domains are able to recognise that they all gain by cooperating. At present, due to the specific agro-ecological constraints and the locally limited role of the state in water resource management, the need for cooperation is higher in the *haor* basin than in other areas. Yet, contrary to what has often been assumed, there is nothing intrinsic to the Bangladeshi socio-cultural system that prevents people from recognising common interests when they exist and organising collectively to pursue them.

Policy Implications

This book challenges some common assumptions about water resource development projects and rural people in Bangladesh.

First of all, the large number and the magnitude of local water management initiatives clearly show that — contrary to what is often assumed — people in rural Bangladesh are not passive recipients or victims of water development projects. If many governmental projects have failed in organising viable water management associations and mobilising local people and resources, the reason probably is not people's incapacity to cooperate. It is more likely to be government agencies' lack of incentives to adopt participatory approaches and mobilise resources locally.

People in rural Bangladesh have the material, technical and organisational capacity to collect, administer and make an optimal use of large amounts of money. Even in cases of large initiatives, they can develop and implement effective strategies to finance water management works. These rapidly developed systems are generally quite flexible; they allow lower income people to provide their labour and higher income people to contribute in cash. Poor people and those who only marginally benefit from an initiative are often exempted from contributing, in particular when the costs involved are not very high. The key to successful mobilisation of material resources is local control, transparency, accountability and a cost effective use of scarce resources. The assumption that local capacity to contribute to O&M is limited or that rural people do not have a 'maintenance culture' needs to be abandoned.

This study has shown that many water development projects delivered by the BWDB basically respond to needs that are strongly felt among all sections of the rural society. However, the top-down approach that characterises the BWDB and its officers' limited contact with the people often prevent this agency from delivering projects that are fine-tuned to specific local needs. Several of the initiatives identified by this study were aimed at adapting BWDB infrastructure to local conditions.

Another important finding is that people's water management interests can transcend differences in socio-economic status. It is therefore possible for socially heterogeneous groups of people to cooperate and for the traditional power elite to have a positive role in water resource management. Conflicting interests between different categories of stakeholders may, of course, exist and have to be carefully assessed by development agencies, when the introduction of new water management technologies in rural areas is being planned. In this domain, applied social scientists need to see beyond class and gender-divisions and to recognise the role of other types of inequalities, such as those between highland and lowland farmers.

Many local initiatives — such as embankment cuts and the construction of crossdams — are well-known phenomena. Embankment cuts in particular often have been either condemned as criminal offences or interpreted as evidence that people are victims of harmful interventions in their environment. In the context of infrastructure rehabilitation projects, the problem of embankment cuts is generally interpreted as a sign of drainage congestion, one to be solved by building sophisticated and costly regulators. The cases presented in this study suggest, however, that embankment cutting is a carefully planned, indigenous water management practice, a much cheaper and often more appropriate one than a regulator. In Chaptir

Haor, where embankment cuts are primarily related to fishing practices and water transport, the costs involved in cutting and closing an embankment are only a fraction of what the yearly maintenance costs of a regulator would be. Though embankment cuts are not in all cases an alternative to regulators, it is essential to collect basic data about the actors, objectives, costs and benefits behind public cuts, before deciding to install costly infrastructure.

Civil engineers often are openly sceptical about people's participation in the design of hydraulic infrastructure. While cutting and closing embankments may not require sophisticated engineering skills, some of the water management infrastructures constructed without any external assistance prove that people in rural Bangladesh have the capacity to understand, design and construct complex water management infrastructure. Good examples are the gravity flow irrigation canals in Chaptir Haor and Chapai-Nawabganj and the drainage outlets on Banskhali's coastal embankment, which differ little in their designs from those generally proposed by the BWDB. A major difference exists, however, in construction cost. Not only were these structures financed by local users themselves, but the work was done at a fraction of the BWDB's usual cost.

The fact that people can and do take initiative to solve their own problems does not mean that they would not welcome external assistance or that they do not consider it as the government's responsibility to provide them with effective water management infrastructure. The many endogenous initiatives do show, however, that external dependence is not an irreversible condition and that it is not as strong as is often assumed. In several study cases people had first tried to get assistance from the government and it was only when they realised that they would not succeed that they mobilised their own resources. BWDB interventions to take over responsibilities that are traditionally taken care of by the people themselves are generally quite welcome, if provided in a timely and appropriate manner.

An analysis of the social organisation of local initiatives can suggest ways of increasing the viability of water management organisations, a recurring issue in debates on people's participation. One sociological finding is that most local initiatives, having a very local impact, are undertaken by people who live and work in the same area, people who already belong to the same social world. Accordingly, they do not need to be organised anew or to meet in formal gatherings to manage effectively their surface water resources and related infrastructure. In many cases decisions to take specific actions are taken in the context of casual but inevitable encounters in fields, markets or mosques or during informal meetings called by the *matabars*. In other cases, such as the *Badh Rakkha Committee*, a powerful, effective and well-respected semi-formal organisation is established. In such organisations traditional rural elites (*matabars*) play important roles. Their local credibility is strong either because they have been elected or because they already are leaders or because of their formal education, management capacity, wealth, and access to government officials. The frequently positive role of the traditional power elite in local initiatives does not mean, however, that they can have an equally positive role in the implementation of exogenous development programmes, where many would not hesitate to grab resources for themselves. Their main incentive to assume a leading role in endogenous initiatives is not to contribute to the development of their area but to protect or enhance their own production system.

Another interesting sociological feature of local initiatives is that women are completely absent, at least from the organisational frontline. Not a single woman was found to be present in meetings related to the planning and monitoring of local water management initiatives or in any of the direct works created, demonstrating that the management of public surface water resources is considered a male domain. However, all women interviewed in the course of this study were fully aware of the water management practices. It is remarkable that, although women are primarily responsible for domestic water supply and men for irrigation, women perceive their water management interests as being basically identical to men's. None of the initiatives had a negative impact on women.

At the central government level there is often scepticism about Union Parishad's capacity to manage local projects. This study has discovered a number of initiatives in which the Union Parishad had a key role, which proves that sometimes Union Parishads have the capacity to — and *de facto* often do — manage small-scale FCD systems. In a number of cases Union Parishad chairmen successfully managed to implement locally identified projects by combining local and external resources (such as Food for Work, wheat), which were acknowledged by many constituents as very useful. This shows that some Union Parishads are aware and responsive to the needs of their constituencies, able to identify and implement relevant water management projects and potentially have the unique capacity to mobilise external as well as internal resources. This is an important conclusion, particularly in the context of the ongoing debate on devolution of control over water management infrastructure and the future role of local government in this sector.

Nevertheless, the government's lingering problems with O&M of public water management infrastructure will not be solved by turning over all responsibilities to the people or to local government agencies. Although more decentralisation of control and authority over the water sector to the local level is essential, people and local level institutions will continue to need professional, material and technical assistance. Endogenous initiatives, however, should no longer be ignored or condemned. They should be closely monitored and kept into account by officials responsible for the nation's water development policies and strategies.

Glossary

Agu	Fixed land lease in cash (local dialect of Banskhali)
Aman	Late monsoon paddy crop, planted in June-July and harvested in November-December
Athi	Homestead land (local dialect of Chaptir Haor)
Aus	Late dry season paddy, planted in March-April and harvested during the monsoon, in June-July
Badh	Embankment
Badhai	First cropping season in Bengal's agricultural calendar, also known as *kharif I*
Badh Rakkha Samity	Embankment Protection Committee
Baishak	Bengali month corresponding to mid-April to mid-May
Bandhak	Mortgage of land or other goods (trees, livestock, cows)
Banna	Damaging, abnormal flood
Bari	Homestead
Barsha	Benevolent, normal flood
Bawa dhan	Broadcast *aman*
Bhaga	Sharecropping of land (local dialect of Galachipa)
Bhanda	Land mortgage (local dialect of Banskhali)
Beel	Low-lying depression in the flood plain which generally retains water throughout the year
Bideshi	Foreign

Bigha	Land measure equal to 33 decimals
Bond	Command area (local dialect of Chaptir Haor)
Borgha	Sharecropping
Boro	Winter season paddy crop, planted in December-January and harvested in April-May
Chaitra	Bengali month corresponding to mid-March to mid-April
Char	temporary or permanent interfluvial island
Chara	Natural stream (local dialect of Banskhali)
Chawkidar	Guard
Chukti	Seasonal land leasing contract (local dialect of Banskhali)
Dawalla	Migrant harvest labourer (local dialect of Chaptir Haor)
Dheki	Wooden mallet used for husking rice
Dhoba	Natural pond, small beel
Dhone	Indigenous irrigation device, also known as *kone*
Ejmali	Collective property
Gamasta	Contractor of *dawallas*, i.e., migrant harvest labourers (local language of Chaptir Hoar)
Gamcha	towel for covering head or shoulders
Goda	Crossdam (local dialect of Banskhali)
Gosti	Patrilineage
Haimantic	Second cropping season in Bengal's agricultural calendar, also known as *kharif II*
Haja	Sharecropping (local dialect of Banskhali)
Haor	Extensive depression between the natural levees of a river, which looks like a saucer, with a deep central part permanently under water
Kanda	Natural river levee (local dialect of Chaptir Haor)

Kani	Land measure equal to 40 decimals
Kechi	Indigenous irrigation device consisting of a swing basket or scoop, also known as *seuni*
Keyar	Land measure equal to 30 decimals
Khaikalashi	Fixed land lease, with rent paid in cash (local dialect of Beel Singri)
Khajna kaura	Land mortgage (local dialect of Beel Singri)
Kalashi	BWDB employee responsible for guarding and operating a regulator
Kamla	Labourer
Kartik	Bengali month corresponding to mid-October to mid-November
Khal	Natural channel or tidal creek
Khara	Small natural creek (local language of Chaptir Haor)
Kone	Indigenous, canoe-shaped irrigation device, made of a hallow tree.
Jali aman	Deep water paddy, also known as 'floating paddy'
Janghal	Contour bundh (local language of Chaptir Haor)
Jaystha	Bengali month corresponding to mid-May to mid-June
Lagit	Fixed land lease in cash (local dialect of Banskhali)
Lungi	Male clothing item. Garment wrapped till ankle-length
Mara nadi	Dead river
Marer kola	Threshing ground (local language of Chaptir Haor)
Matabar	Informal leader of a village or neighbourhood
Maund	Weight measure equal to 37 kg
Nadi	River
Nasi	Drainage pipe (local language of Chittagong)
Palla	Irrigation canal (local language of Chaptir Haor)

Panchayat fund	Collective village funds found in Chaptir Haor
Patit jami	High, fallow land used for grazing cattle (local dialect of Chaptir Haor)
Patwari	Tax collector
Phalgun	Bengali month corresponding to mid-February to mid-March
Pradhan	Leader
Rabi	Dry cropping season (October to May)
Salish	Informal village court
Samaj	Society, also refers to indigenous social institution through which villagers enact and enforce normative rules
Sangha	Flood resistant grassy plant commonly used in low land areas for demarcating plots
Seuni	Indigenous irrigation device, also known as *kechi*
Shail	Transplanted Aman
Taka	Money, also the name of Bangladesh currency

1. A floodplain in summer

2. A floodplain in winter

3. Rice fields in south-east Bangladesh

4. The Mahananda River in Chapai Nawabganj

5. On the way to Galachipa

6. On the way to Galachipa, another scene

7. View from the river in Galachipa

8. Another view from the river in Galachipa

9. View from the river in Galachipa

10. Jute, Bangladesh's "golden fibre"

11. Jute collected for loading

12. Levelling land with the help of a wooden beam

13. Transplantation of young paddy plants

14. Harvesting *boro* paddy

15. Cow dung dried for fuel

16. Traditional *rabi* crop

17. Inland open water fishery

18. Inland open water fishery, another scene

19. *Kechi*, a traditional irrigation device

20. *Kone*, another traditional irrigation device

21. A low-lift pump

22. Irrigation canal in Galachipa

23. A BWDB five-went regulator

24. Crossdam for flood protection in Chaptir Haor

25. A *goda* in Banskhali's hilly area retains water for horticulture

26. Small crossdam in Banskhali

27. Small crossdam in Banskhali

28. The Bamerchara Irrigation Scheme

29. A crossdam for fishery in Banskhali

30. Drainage outlets with earth piles to sustain fishing gear (Banskhali)

31. Partition dike with salt fields (left) and paddy (right)

32. A shrimp inlet

33. "Maintenance" of a leaking regulator

34. A “naked” homestead in Chaptir Haor

35. A wealthy farmer's homestead in Chaptir Haor

36. "*Patit jami*" used as threshing ground (Chaptir Haor)

37. Temporary "*dawalla*" shelter in Chaptir Haor

38. A "*dawalla*" boat in Chaptir Haor

39. Pots to be exchanged with paddy in Chaptir Haor

40. The approach of a hailstorm in Chaptir Haor

Bibliography

Adnan, S., 1988, *Village Studies*, Comilla: BARD.

———, 1991, *Floods, People and the Environment*, Dhaka: Research and Advisory Services.

———, 1999, 'Agrarian Structure and Agricultural Growth Trends in Bangladesh: The Political Economy of Technological Change and Policy Interventions.' In: B. Rogaly *et al.*, (ed.), *op.cit.*

———, *et al.*, 1992, *People's Participation, NGOs and the Flood Action Plan. An Independent Review*, Dhaka: RAS/Oxfam.

Ahmad, M., 1989, 'Deluge in the Delta', In: I. Ahmad (ed.), *Flood in Bangladesh*, Dhaka: Community Development Library.

Ahmed, I. (ed.), 1999, *Living with Floods: An Experience of Alternatives*, Dhaka: University Press Ltd.

Ahmed, M., 1997, 'Socio-economic and Policy Issues in the Floodplain Fisheries of Bangladesh', In: Tsai and Ali, *op. cit.*

Ahmed, Z.U., 1982, *Bizna. A Study of Power Structure in Contemporary Rural Bangladesh*, Dhaka: Contemporary Books Int.

Ahmed, B., and M. Rahman, 1990, 'Local Organizations and Adaptations of Irrigation Technology', In: S. Huq, A.A. Rahman and G.R. Conway, *op.cit.*

Ali, M.Y., 1996, *Fish, Water and People. Reflections on Inland Openwater Fisheries Resources of Bangladesh*, Dhaka: University Press Ltd.

Ali, S.I., 1990, 'Haor Basin Eco-System', In: A.A. Rahman, S. Huq, and G.R. Conway, *op.cit.*

Ambler, J.S., 1993, 'Rethinking the Language of Farmer Water Users Association', *FMIS Newsletter*, Colombo: IIMI.

Arens, J. and J. van Beurden, 1977, *Poor Peasants and Women in a Village in Bangladesh*, Amsterdam: Third World Publ.

Bagadion, B.U., and F.K. Korten, 1985, 'Developing Irrigators' Organisations: Learning Process Approach to Participatory Irrigation Programmes', In: M. Cernea (ed.), *op. cit.*, 52-90.

Bardhan, Pranab, and J. Dayton-Johnson, 2000, 'Heterogeneity and Commons Management', Paper presented at the 8th Biennial Conference of the International Association for the Study of Common Property (IASCP).

Barenstein, J., 1994a, *Overcoming Fuzzy Governance in Bangladesh. Policy Implementation in Least Developed Countries*, Dhaka: University Press Ltd.

———, 1994b, 'The fiscal misery of local government in Bangladesh', *The Journal of Social Studies,* 66: 97-105.

———, 2000a, *Local governance and civil society in the 1990s — A rapid empirical probe in Bangladesh*, Background document for the Bangladesh National Institutional Review (BNIR), Washington, D.C.: The World Bank.

———, 2000b, *Towards Local Governance in Bangladesh – From Failed Paradigms to Alternative Interpretations*, PhD Thesis, Manchester: University of Manchester, Faculty of Economic and Social Studies, Institute for Development Policy and Management (IDPM).

Bertocci, P.J., 1970, *Elusive Villages: Social Structure and Community Organization in Rural East Pakistan*, Michigan: Michigan University Press.

———, 1976, 'Social Organization and Agricultural Development in Bangladesh', In: Stevens, R.D. *et al.* (eds.), *Rural Development in Bangladesh and Pakistan,* Honolulu: University Press of Hawaii.

Blair, H.W., 1974, *The Elusiveness of Equity. Institutional Approaches to Rural Development in Bangladesh*, New York: Centre for International Studies.

———, 1978, 'Rural Development, Class Structure and Bureaucracy in Bangladesh', *World Development* 6/1: 65-82.

Blair, H.W. (ed.), 1989, *Can Rural Development be Financed from Below? Local Resource Mobilization in Bangladesh*, Dhaka: University Press Ltd.

Bosc, P.M. and Weber J., 2000, *Farmers' Organizations and the Building of a New Common Pool Resource: the Role of Social Heterogeneity in the Socio-Political Search of Compromises*, Paper presented at the 8th Biennial Conference of the International Association for the Study of Common Property (IASCP).

Botrall, A., 1983, *Review of Irrigation Management Practices in Bangladesh*, Dhaka: Centre for Development Science, BUP.

Boyce, J.K., 1987, *Agrarian Impasse in Bengal. Institutional Constraints to Technological Change,* New York: Oxford University Press.

———, 1990, 'Birth of a megaproject: political economy of flood protection in Bangladesh', *Environmental Management*, 14(4): 419-28.

———, 1990a, 'Floods in Bangladesh I: Geographical Background to the 1987 and 1988 floods', *The Geographical Journal*, 156(2): 12-22.

———, 1990b, 'Floods in Bangladesh II. Flood mitigation and environmental aspects', *The Geographical Journal*, 156(2): 156-65.

Bromley, D.W., (ed.), 1992, *Making the Commons Work. Theory, Practice and Policy*, San Francisco: ICS Press.

Bromley, D.W., and M.M. Cernea, 1989, *The Management of Common Property Natural Resources — Some Conceptual and Operational Fallacies,* Washington D.C.: The World Bank.

Bromley, D.W. and Chapagain, 1984, 'The Village Against the Centre. Resource Depletion and in South Asia', *American Journal of Agricultural Economics,* 66: 868-73

Bruns, B., 1993, 'Promoting Participation in Irrigation: Reflections on Experience in Southeast Asia', *World Development*, 21(11): 1837-49.

BWDB/SRP, 1992, *Benchmark and Evaluation Study. Benchmark Study Report,* Technical Report 25, Dhaka: BWDB.

———, 1994a, *Feasibility Report. Chaptir Haor Sub-Project,* Dhaka: BWDB.

———, 1994b, *Feasibility Report. Beel Singri Sub-Project*, Dhaka: BWDB.

Cernea, M.M. (ed.), 1985, *Putting People First: Sociological Variables in Rural Development*, New York: OUP.

Chowdhury, Y.A., 1988, 'Institutional Analysis: BWDB and BADC', In: *Bangladesh Agriculture Sector Review, Vol: V: Institutions*, Dhaka: UNDP, pp. 1-60.

Coward, W.E., 1979, 'Principles of Social Organization in an Indigenous Irrigation System', *Human Organization* 38: 28-36.

———, 1985, "Technical and Social Change in Currently Irrigated Regions: Rules, Roles and Rehabilitation", In: M. Cernea, *op. cit*.

Coward, W.E. (ed.), 1980, *Irrigation and Agricultural Development in Asia. Perspectives from the Social Sciences*, Ithaca: Cornell UP.

Downing and Gibson (ed.), 1974, *Irrigation's Impact on Society.*

de Vylder, S., 1982, *Agriculture in Chains*, London: Zed Press.

Duyne, J.E., 1987, *Small Credits and Development. The Influence of socio-economic factors on poor oriented credit programmes in rural Bangladesh*, M. A. Thesis, Zurich: Department of Social Anthropology, University of Zurich.

———, 1993, *The Socio-economic Dimensions of Water Management Project,* Technical Report 36. Dhaka: BWDB/SRP.

———, 1994a, *Ushering to Participation: Current attempts to mobilise farmers in large scale water management systems in Bangladesh,* Paper presented to the 1994 Workshop of the European Network of Bangladesh Studies, Woudschoten (The Netherlands), August 1994.

———, 1994b, *Socio-Economic Determinants of Participatory Water Management and Cost Recovery. A Study of the Karnaphuli Irrigation Project*, Technical Report 37, Dhaka: BWDB/SRP.

———, 1994c, *Embankment Maintenance Groups. An assessment of the social, economic and institutional implications of women's involvement in rural infrastructure maintenance programmes*, Technical Report no. 43, Dhaka: BWDB/SRP.

———, 1995b, *Towards Participatory Water Management. Some Practical Suggestions on How to Enhance People's Participation with special reference to flood control and drainage systems,* Technical Report no. 49, Dhaka: SRP/BWDB.

———, 1995c. 'Gender Issues in Water Management', In: *Training Manual on Integrated Water Resources Management*, Delft and Dhaka: BUET-DUT University Linkage Project.

———, 1997a, *An Empirical Assessment of the Role of Water Users Organisations in Participatory Water Management,* Technical Report no. 54, Dhaka: SRP/BWDB.

———, 1997b, 'The Process to Revise the Guidelines for People's Participation in Water Development Projects', *The Daily Star,* Dhaka, December 18, 1997.

———, 1997c, 'People's Initiatives in Water Management', *The Daily Star*, Dhaka, December 18, 1997.

———, 1998, 'Local Initiatives: People's Water Management Practices in Rural Bangladesh', *Development Policy Review,* 16(3): 265-80.

———, 1999a, 'Local Initiatives and National Projects,' In: A.K. Datta (ed.), *Planning and Management of Water Resources*, Dhaka: University Press Ltd, pp. 75-90.

———, 1999b, 'Introducing Participatory Management in Flood Control, Drainage and Irrigation Systems', In: H.P. Middendorp *et al.*, *Sustainable Inland Fisheries Management in Bangladesh,* Manila: ICLARM.

Edwards, V. M., 2000, *The Role of Contextual Factors in Common Pool Resource Analysis*, Paper presented at the 8th Biennial Conference of the International Association for the Study of Common Property (IASCP).

Esman, M.J. and N. Uphoff, 1984, *Local Organizations. Intermediaries in Rural Development*, Ithaca: Cornell UP.

FAP 6, 1993, *Local Initiatives and People's Participation in the Management of Water Resources,* Dhaka: FPCO/Northwest Hydraulic Consultants.

Faraizi, A.H., 1994, *Bangladesh. Peasant Migration and the World Capitalist Economy*, Dhaka, University Press Ltd.

Farooque, M, 1999, 'Analysis of the Legal Regime for Inland Fisheries in Bangladesh', In: H.P. Middendorp *et al.*, *Sustainable Inland Fisheries Management in Bangladesh,* Manila: ICLARM.

Feeny, D., *et al.* (1990), 'The Tragedy of the Commons: Twenty-Two Years Later', *Human Ecology* 18(1): 1-19.

Freeman, D.M., 1989, *Local Organizations for Social Development. Concepts and Cases of Irrigation Organization.* Boulder: Westview Studies.

Gardner, K., 1995, *Global Migrants, Local Lives: Travel and Transformation in Rural Bangladesh,* Oxford: Clarendon Press.

Geertz, C., 1972, 'The Wet and the Dry: Traditional Irrigation in Bali and Morocco', *Human Ecology*, 1:23-39.

Gisselquist, D., 1991, *Medium and Large Surface Water Development in Bangladesh. Analysis of Failures and Recommendations for Reform*, Unpublished report prepared for the World Bank, Dhaka: The World Bank Regional Office.

GOB/BBS, 1989, *The Bangladesh Census of Agriculture and Livestock: 1983-1984. Zila Series: Chittagong, Chapai-Nawabganj, Pathuakali and Sunamganj,* Dhaka: Bangladesh Bureau of Statistics.

GOB/BBS, 1992, *Population Census 1991. Zila Series: Chittagong, Chapai-Nawabganj, Pathuakali and Sunamganj,* Dhaka: Bangladesh Bureau of Statistics.

GOB/BBS (n.d), 'The Bangladesh Census of Agriculture and Livestock: 1995-1996', Unpublished document, Dhaka: Bangladesh Bureau of Statistics.

GOB/Planning Commission, 1990, *The Fourth Five Year Plan (1990 95),* Dhaka: Government Press.

GOB/Ministry of Water Resources, 1995, *The Guidelines for People's Participation*, Dhaka: MOWR.

Guimareas, J.P.C., 1989, 'Shrimp culture and market incorporation: A study of shrimp culture in paddy fields in the southwest of Bangladesh', *Development and Change*, 20(4): 653-82.

Habib, I., 1963, *The Agrarian System of Mogul India, 1556-1707*, London: Asia Publishing House.

Haggart, K., *et al.* (eds.), 1994, *Rivers of Life*, London and Dhaka: Panos/ BCAS.

Hall, A., 1988, 'Community Participation and Development Policy: A Sociological Perspective', In: A. Hall and J. Midgley (eds), *Development Policies. Sociological Perspectives,* Manchester: Manchester University Press.

Hanchett, S., 1997, 'Participation and Policy Development. The case of the Flood Action Plan', *Policy Development Review*, 15: 277-95.

Haque, C.E., 1997, *Hazards in a Fickle Environment: Bangladesh*, Dordrecht: Kluwer Academic Publisher.

Hardin, G., 1968, 'The Tragedy of the Commons', *Science*, 162: 1243-48.

Hardin, R., 1982, *Collective Action,* Baltimore, Johns Hopkins UP.

Hartmann, B. and J.K. Boyce, 1983, *A Quite Violence: View from a Bangladesh Village*, London: Zed Press.

Heckathorn, D.D., 1993, 'Collective Action and Group Heterogeneity: Voluntary Provision Versus Selective Incentives', *American Sociological Review*, 58: 329-350.

Hofer, T., 1998, *Floods in Bangladesh: A Highland-Lowland Interaction?,* Bern: Geographica Bernensia Series.

Howes, M., 1985, *Whose Water? An Investigation of the Consequences of Alternative Approaches to Small Scale Irrigation in Bangladesh,* Dhaka: BIDS.

Hughes, R.S., *et al.,* 1994, *Floodplains or Floodplans? A Review of Approaches to Water Management in Bangladesh,* London and Dhaka: IIED and RAS.

Hunt, R., and E. Hunt., 1976, 'Canal Irrigation and Local Social Organization', *Current Anthropology*, 17: 389-411.

Huq, S., A.A. Rahman and G.R. Conway, 1990 (eds.), *Environmental Aspects of Agricultural Development in Bangladesh*, Dhaka: University Press Ltd.

International Irrigation Management Institute (IIMI), 1995, *Workshop on the Study of Privatization of Minor Irrigation in Bangladesh. vol. I: Papers and Proceedings,* Dhaka: IIMI/BARC.

Islam, A., 1989, 'A Review of Some Aspects of Fisheries Sub-sector', In: UNDP (ed.), *Bangladesh Agriculture Sector Review, vol: II: Sub-Sectors of Agriculture.* Dhaka: UNDP.

Islam, S., 1985, *Villages in the Haor-Basin of Bangladesh. Studies in Socio-Cultural Change In Rural Bangladesh*, Tokyo: University of Foreign Studies, Institute for the Study of Languages and Cultures of Asia and Africa.

Islam, M.A., 1990, 'The Alluvial Plains of Bangladesh', In: Rahman A.A., S. Huq and G.R. Convey, *op.cit.,* pp. 47-51.

Jahan, I. and Sanyal K.K., 1993, *1990–91 Rabi/Boro Season*, Dhaka: International Fertilizer Development Centre.

Jahangir, B.K., 1979, *Differentiation, Polarisation and Confrontation in Rural Bangladesh*, Dhaka: CSS.

Jansen, E.G., 1987, *Rural Bangladesh. Competition for Scarce Resources,* Dhaka: University Press Ltd.

Januzzi, F. T. and J.T. Peach, 1980, *The Agrarian Structure of Bangladesh: An Impediment for Development,* Colorado: Westview Press.

Jodha, N.S., 1986, 'Common Property Resources and Rural Poor in Dry Regions of India', *Economic and Political Weekly*, 21:1169-81.

Johanson J. and K. Hultin, 1992, *Sluice Committees in Perspective. A study on Sluice Committee Organisation in Polder 43/2C, Pathuakali District, Bangladesh,* Stockholm: Department of Social Anthropology (Development Studies Unit), Working Paper no. 21.

Khan, H.R., 1990, 'Water Control and Productivity', In: Huq, S., A.A. Rahman and G.R. Conway, 1990, Huq, S., A.A. Rahman and G.R. Conway, *op.cit*, pp. 127-34.

Khan, M.A., 1997, 'Ecology of Floodplains in the Northeastern Region of Bangladesh', In: Tsai and Ali, *op. cit.*

Khan, M.S., *et al.*, 1994, *Wetlands of Bangladesh*, Dhaka: Bangladesh Centre for Advanced Studies (BCAS) and Nature Conservation Movement.

Korten, D., 1980, 'Community Organization and Rural Development. A Learning Process Approach', *Public Administration Review*, 40: 480-511.

Lansing, J.S., 1989, "Balinese 'Water Temples' and the Management of Irrigation", *American Anthropologist*, 89: 326-41.

Leach, E.R., 1959, 'Hydraulic Society in Ceylon,' *Past and Present*, 15: 2-25.

Li, Tania Murray, 1996, 'Images of Community. Discourse and Strategy in Property Relations', *Development and Change*, 27: 501-27.

Lindquist, A.C., 1989, 'Project Aid Agriculture: Major (Surface Water) Flood Control, Drainage and Irrigation Projects', In: UNDP: *Bangladesh Agriculture Sector Review, vol: III: Land, Water and Irrigation*, Dhaka: UNDP.

Mandal, M.A.A., 1989, 'Market for and Returns from Groundwater Irrigation in Bangladesh', In: UNDP: *Bangladesh Agriculture Sector Review, vol: III: Land, Water and Irrigation*, Dhaka: UNDP.

Marwell, G., *et al.*, 1988, 'Social Networks and Collective Action. A Theory of the Critical Mass', *American Journal of Sociology*, 94(3): 502-34.

Marwell, G., and P. Oliver, 1993, *The Critical Mass in Collective Action. A Micro-Social Theory,* Cambridge: Cambridge University Press.

Mayoux, L., 1995, 'Beyond Naivety: Women, Gender Inequality and Participatory Development', *Development and Change*, 26(2): 235-58.

McCay, B.J., and M. Acheson, (eds.), 1987, *The Question of the Commons: The Culture and Ecology of Communal Resources,* Tucson: University Press of Arizona.

McKean M.A., 1992, 'Success of the Commons. A Comparative Examination of Institutions for Common Property Resource Management', *Journal of Theoretical Politics,* 4(3): 247-81.

Mearns, R., 1995, 'Community, Collective Action and Common Grazing: The Case of Post-Socialist Mangolia', *The Journal of Development Studies*, 32(3):297-339.

Meinzen-Dick, R., *et al.*, 1995, *Participation in Irrigation*, Participation Series no. 3, Environment Department Papers, Washington D.C.: The World Bank.

Melucci, A., 1995, *Challenging Codes. Collective Action in the Information Age.* Cambridge: Cambridge UP.

Middendorp, H.A.J., P. Thompson and R.S. Pomeroy (eds.), 1999, *Sustainable Inland Fisheries Management in Bangladesh,* Manila: ICLARM Publications.

Minkin, S.F., 1989, *Steps for conserving and developing Bangladesh fish resources,* Dhaka: UNDP.

Montgomery, R., 1985, 'The Bangladesh floods of 1984 in historical context', *Disasters*, 9(3): 163-172.

Mukherjee, R., 1971, *Six Villages of Bengal*, Bombay: Popular Prakashan.

Nagel, J.H., 1987, *Participation*, Englewood Cliffs (New Jersey): Prentice-Hall.

Netting, R., 1981, *Balancing on an Alp*, Cambridge: Cambridge University Press.

Nishat, A., 1989, 'Large Scale Water Development in Bangladesh', In: UNDP: *Bangladesh Agriculture Sector Review, vol: III: Land, Water and Irrigation*, Dhaka: UNDP.

North, D.C., 1990, *Institutions, Institutional Change and Economic Performance*, Cambridge: Cambridge University Press.

Oakerson, R.J., 1992, 'Analyzing the Commons: A Framework', In: D. Bromley, *op. cit.*

Olson, M., 1965, *The Logic of Collective Action*, Cambridge (Mass.): Harward UP.

Oliver, P., and G. Marwell, 1985, *A Theory of the Critical Mass. Interdependence, Group Heterogeneity and the Production of Collective Action.*

Ostrom. Elinor, 1985, *'Are Successful Efforts to Manage Common-pool Problems a Challange to the Theories of Garrett Hardin and Mancur Olson?'*, Working Paper, Workshop in Political Theory and Policy Analysis, Bloomington: Indiana University.

———, 1990, *Governing the Commons: The Evolution of Institutions for Collective Action*, New York: Cambridge University Press.

———, 1992a, *Crafting Institutions for Self-Governing Irrigation Systems,* San Francisco: ICS Press.

———, 1992b, 'The Rudiments of a Theory of the Origins, Survival, and Performance of Common Property Institutions', In: D. Bromley, *op. cit.*

Ostrom, E., 1992c, 'Community and the Endogenous Solution of Commons Problems', *Journal of Theoretical Politics*, 4(3): 343-52.

Ostrom, Elinor, *et al.*, 1993, *Institutional Incentives and Sustainable Development. Infrastructure Policies in Perspective*, Boulder: Westview Press.

Oversee Development Agency (UK), 1997, 'Impacts of Flood Control and Drainage With or Without Irrigation (FCD/I) Project on Fish Resources and Fishing Communities in Bangladesh: Executive Summary of the FAP 17 Final Report,' In: Tsai and Ali (eds.), *op. cit.*

Palmer-Jones, R., 1999, 'Slowdown in Agricultural Growth in Bangladesh: Neither a Good Description nor a Description Good to give', In: B. Rogaly *et al.* (eds.), *op. cit.*

Parlin. B.W., and M.W. Lusk (eds.), 1991, *Farmer Participation and Irrigation Organization,* Boulder: Westview Press.

Paul, B.K., 1984, 'Perception of and agricultural adjustments to floods in Jamuna Floodplain, Bangladesh', *Human Ecology,* 12: 3-19.

Paul, B.K., 1995, 'Farmers' Responses to the Flood Action Plan (FAP) of Bangladesh: An empirical Study', *World Development*, 23(2): 299-309.

Paul, N.C. 1997, 'Openwater Fisheries in the Northeast Region of Bangladesh', In: Tsai and Ali, *op. cit.*

Pitman, K.G.T., 1993, 'National Water Planning in Bangladesh 1985-2005. The role of groundwater in irrigation development', In: F. Kahnert and G. Levine (eds.), *Groundwater Irrigation and the Rural Poor: Options for Development in the Gangetic Basin: A World Bank Symposium*, The World Bank: Washington DC.

Pokrant, B., P. Reeves and J. McGuire, 1997, 'Riparian Rights, the Organisation of Work and Market Relations among the Inland Fishers of Colonial Bengal, 1793 to 1950', In: Tsai and Ali, *op. cit.*

Popkin, S., 1979, *The Rational Peasant. The Political Economy of Rural Society in Vietnam,* Berkeley: University of California Press.

———, 1993, 'Public Choice and Peasant Organization', In: C.R. Russel and N. K. Nicholson, *Public Choice and Rural Development,* Chicago: Chicago University Press.

Rahman, A., 1995, *Beel Dakatia. The Environmental Consequences of a Development Disaster,* Dhaka: Universty Press Ltd.

Rahman, A.A.K., 1997, 'Fish Marketing in Bangladesh: Status and Issues', In: Tsai and Ali, *op. cit.*

Rahman, H.Z., 1989, *Critical Review of Studies on Organisation of Production,* Dhaka: BIDS.

———, 1994, *Re-thinking the Land Reform Question: Report From an Inter-Face Study,* Dhaka: BIDS.

———, 1998, *Poverty Issues in Bangladesh. A Strategic Review*, Study Commissioned by DFID, Dhaka.

Rahnema, M., 1992, 'Participation', In: W. Sachs (ed.), *The Development Dictionary: A Guide to Knowledge as Power,* London: Zed Books.

Rahman, A.A., S. Huq and G.R. Convey (eds.), 1990a, *Environmental Aspects of Surface Water Systems of Bangladesh,* Dhaka, University Press Ltd.

———, 1990b, *Environmental Aspects of Agricultural Development in Bangladesh*, Dhaka, University Press Ltd.

Rashid H.E., 1989, 'Land Use in Bangladesh: Selected Topics', In: UNDP (ed), *Bangladesh Agriculture Sector Review, vol: III: Land, Water and Irrigation,* Dhaka: UNDP.

Rashid, H.E., 1991, *Geography of Bangladesh,* Dhaka: University Press Ltd.

Rashid, H, and B.K. Paul, 1987, 'Floods in Bangladesh. Is there an indigenous solution?', *Environmental Management*, 11/2: 155-73.

Rasid, H., 1993, 'Preventing flooding or regulating flood levels? Case studies on Perception of flood alleviation in Bangladesh', *Natural Hazards*, 8: 39-57.

Repetto, R., 1986, *Skimming the Water: Rent-seeking and the Performance of Public Irrigation Systems,* Research Report 4, Washington DC: The World Resource Institute.

Rizvi, S.N.H., 1970, *East Pakistan District Gazetteers. Sylhet,* Dacca: Government Press.

———, 1975, *Bangladesh District Gazetteers. Chittagong,* Dacca: Government Press

Rogaly, B., B. Harriss-White and S. Bose (eds.), 1999, *Sonar Bangla? Agricultural Growth and Agrarian Change in West Bengal and Bangladesh,* London: Sage.

Rogaly, B., B. Harriss-White and S. Bose, 1999, 'Introduction: Agricultural Growth and Agrarian Change in West Bengal and Bangladesh', In: B. Rogaly *et al., op. cit.*

———, 1992, 'Common Property and Collective Action in Economic Development', In: Bromley (ed.), *op. cit.*

Saha, B.K., 1997, *Agrarian Structure and Productivity in Bangladesh and West Bengal. A Study in Comparative Perspective,* Dhaka: University Press Ltd.

Saleh A.F.M. and A. Nishat, 1989, 'Review of Assessment of Ground Water Resources: Quality and Quantity', In: UNDP (ed.), *Bangladesh Agricultural Sector Review, vol. 3: Land, Water and Irrigation*, Dhaka: UNDP.

Stalker, P., 1997, *The Dancing Horizon. Human Development Prospects for Bangladesh*, Dhaka: United Nations Organisations.

Sadeque, S.Z., 1992, *Environment and Natural Resource Management in Bangladesh*, Dhaka: BSA.

Schlager E. and W. Blomquist, 1998, 'Heterogeneity and Common Pool Resource Management', In: E.T. Loehman and D.M. Kilgour (eds.), *Designing Institutions for Environmental and Resource Management,* Cheltenham (UK): Edward Elgar Press.

Schmuck-Wildmann H., 1996, *Living with the Floods. Survival Strategies of Char Dwellers in Bangladesh,* Berlin: ASA Verlag.

Siddique, A., 1976, *Bangladesh District Gazetteers. Rajshahi,* Dacca: Bangladesh Government Press.

Singleton, S. and Taylor, M., 1992, 'Common Property, Collective Action and Community', *Journal of Theoretical Politics* 4(3):309-24.

Small, L. E., *et al.*, 1989, *Financing Irrigation Services. A Literature Review and Selected Case Studies from Asia,* Colombo (Sri Lanka): International Irrigation Management Institute.

Soussan, J. and A. Datta, (eds.), 1998, *Community Partnership for Sustainable Water Management,* Dhaka: University Press Ltd.

Soussan, *et al.*, 1998, 'Social and Economic Impacts', In: J. Soussan and A. Datta, *op. cit.*, vol. 2.

Taylor, M., 1987, *The Possibility of Cooperation*, Cambridge: Cambridge University Press.

Taylor, M., and S. Singleton, 1993, 'The Communal Resource: Transaction Costs and the Solution to Collective Action Problems', *Politics and Society*, vol. 21(2):195-214.

Thompson, P.M and P. Sultana, 1996a, 'Distributional and Social Impacts of Flood Control in Bangladesh', *The Geographical Journal*, vol. 162(1): 1-13.

———, 1996b, 'Operation and Maintenance Performance and Conflicts in Flood-control Projects in Bangladesh', *Water Resource Development*, vol. 12(3): 311-28.

Toufique K.A., 1999, 'Property Rights and Power Structure in Inland Fisheries in Bangladesh', In: H.A.J. Middendorp *et al.*, *Sustainable Inland Fisheries Management in Bangladesh*, Manila: ICLARM.

Tsai Chu-fa and M. Y. Ali, 1997, *Openwater Fisheries of Bangladesh*, Dhaka: University Press Ltd.

Ullah, M., 1996, *Land, Livelihood and Change in Rural Bangladesh,* Dhaka: University Press Ltd.

UNDP, 1994, *UNDP's 1994 Report of Human Development in Bangladesh: Empowerment of Women*, Dhaka: UNDP.

Uphoff, N. (ed.), 1986, *Local Institutional Development: An Analytical Sourcebook with Cases,* West Hartford: Kumarian Press.

Uphoff, N., 1986a, *Improving International Irrigation Management with Farmer Participation: Getting the Process Right,* Boulder: Westview Press.

Van Schendel, W., 1982, *Peasant Mobility. The Odds of Life in Rural Bangladesh,* New Delhi: Manohar Publisher.

Vedeld, T., 2000, *Village Politics: Heterogeneity, Leadership and Collective Action,* Paper presented at the 8th Biennial Conference of the International Association for the Study of Common Property (IASCP).

Wade, Robert, 1979, 'The Social Response to Irrigation: An Indian Case Study', *Journal of Development Studies* 16(1): 3-26.

———, 1982, 'The System of Political and Administrative Corruption: Canal Irrigation in South India', *The Journal of Development Studies*, 18(3):287-325.

———, 1988, *Village Republics. Economic Conditions for Collective Action in South India,* Cambridge: Cambridge University Press.

———, 1992, 'Common-Property Resource Management in South Indian Villages', In: D. Bromley, *op. cit.*

White, S., 1984, *Arguing with the Crocodile: Gender and Class in Bangladesh,* Dhaka: University Press Ltd.

Willocks, W., 1930, *Lectures on the Ancient Systems of Irrigation in Bengal,* Calcutta: University of Calcutta.

Wittfogel, K.A., 1957, *Oriental Despotism. A Comparative Study of Total Power,* Yale: Yale University Press Inc.

Wood, G, 1995, *A Review of the Guidelines for People's Participation in Water Development Projects* (draft), Dhaka: UNDP.

———, 1996, *Playing with Rivers: Democracy and the Water Sector in Bangladesh,* Paper prepared for the Second Conference on the Flood Action Plan, Brussels: European Parliament.

World Bank, 1970, *Proposal for an Action Programme: East Pakistan agriculture and water development.* vol. 3, *The Water Program*, Report no. PS-2a, Washington DC: The World Bank.

———, 1972, *Bangladesh: Land and Water resources sector study,* vol. 8, *The Flood Problem*, Technical Report no. 24, Washington DC: The World Bank.

———, 1987, *Bangladesh: Second small scale flood control, drainage and irrigation project,* Report no. 6904 BD, Washington DC: The World Bank.

———, 1990, *Performance Audit Report: Bangladesh Drainage and Flood Control Project,* Report no. 8805, Asian Region, Washington DC: The World Bank.

———, 1991, *Bangladesh Fisheries Sector Review,* Report no. 8830-BD. Washington D.C: The World Bank.

———, 1996a, *The World Bank Participation Sourcebook*, Washington, DC: The World Bank.

———, 1996b, *From Vision to Action in the Rural Sector,* Washington DC, The World Bank.

———, 1998, *Bangladesh. From Counting the Poor to Making the Poor Count,* Washington DC: The World Bank.

———, 1999, *Bangladesh. Key Challenges for the next Millennium,* Dhaka: The World Bank.

Index